T0322015

Complexity and the Arrow of Time

There is a widespread assumption that the universe in general, and
life in particular, is "getting more complex with time." This book
brings together a wide range of experts in science, philosophy, and
theology and unveils their joint effort in exploring this idea. They
confront essential problems behind the theory of complexity and the
role of life within it. What is complexity? When does it increase, and
why? Is the universe evolving towards states of ever greater
complexity and diversity? If so, what is the source of this universal
enrichment? This book addresses those difficult questions, and
offers a unique cross-disciplinary perspective on some of the most
profound issues at the heart of science and philosophy. Readers will
gain insights into complexity that reach deep into key areas of
physics, biology, complexity science, philosophy, and
religion.

Complexity and the Arrow of Time

Edited by

CHARLES H. LINEWEAVER
Australian National University

PAUL C. W. DAVIES
Arizona State University

MICHAEL RUSE
Florida State University

CAMBRIDGE
UNIVERSITY PRESS

Shaftesbury Road, Cambridge CB2 8EA, United Kingdom

One Liberty Plaza, 20th Floor, New York, NY 10006, USA

477 Williamstown Road, Port Melbourne, VIC 3207, Australia

314–321, 3rd Floor, Plot 3, Splendor Forum, Jasola District Centre, New Delhi – 110025, India

103 Penang Road, #05–06/07, Visioncrest Commercial, Singapore 238467

Cambridge University Press is part of Cambridge University Press & Assessment, a department of the University of Cambridge.

We share the University's mission to contribute to society through the pursuit of education, learning and research at the highest international levels of excellence.

www.cambridge.org
Information on this title: www.cambridge.org/9781107027251

© Cambridge University Press & Assessment 2013

This publication is in copyright. Subject to statutory exception and to the provisions of relevant collective licensing agreements, no reproduction of any part may take place without the written permission of Cambridge University Press & Assessment.

First published 2013

A catalogue record for this publication is available from the British Library

Library of Congress Cataloging-in-Publication data
Complexity and the arrow of time / edited by Charles H. Lineweaver,
Paul C. W. Davies and Michael Ruse.
    pages cm
Includes bibliographical references and index.
ISBN 978-1-107-02725-1 (hardback)
1. Complexity (Philosophy)   2. Science – Philosophy.   I. Lineweaver,
C. H. (Charles H.)   II. Davies, P. C. W.   III. Ruse, Michael.
Q175.32.C65C64   2013
003 – dc23   2013007433

ISBN   978-1-107-02725-1   Hardback

Additional resources for this publication at www.cambridge.org/9781107027251

Cambridge University Press & Assessment has no responsibility for the persistence or accuracy of URLs for external or third-party internet websites referred to in this publication and does not guarantee that any content on such websites is, or will remain, accurate or appropriate.

Contents

Author biographies

ERIC J. CHAISSON is a Research Associate at the Harvard-Smithsonian Center for Astrophysics and a member of the Faculty of Arts and Sciences at Harvard, where he teaches an annual undergraduate class on "cosmic evolution". His scientific research involves an interdisciplinary, thermodynamic study of physical and biological phenomena that seeks to understand the origin, evolution, and unification of galaxies, stars, planets, and life forms in the universe. His educational research engages experienced teachers and computer animators in creating better methods, technological aids, and novel curricula to stimulate teachers and instruct students in all aspects of the natural sciences. He is currently a board member of the International Big History Association.

PHILIP CLAYTON is Ingraham Professor and Dean of the Faculty at Claremont School of Theology and Provost of Claremont Lincoln University. He received a joint doctorate in philosophy and religious studies from Yale University and has held teaching posts at Williams College and the California State University, as well as guest professorships at the University of Munich, the University of Cambridge, and Harvard Divinity School. The author or editor of some two dozen books, Clayton specializes in the philosophy of science, especially theoretical biology and emergent complexity, and in the philosophy of religion and comparative religious studies.

SIMON CONWAY MORRIS is professor of evolutionary paleobiology at Cambridge University and a Fellow of St. John's College. He was elected to the Royal Society in 1990 and has won various medals and awards. His contribution to the Burgess Shale was summarized in *The Crucible of Creation*, while some of his earlier work on evolutionary convergence is discussed in *Life's Solution* (Cambridge University Press). He contributes to the public understanding of science and the science/religion debates.

PAUL C. W. DAVIES is a Regents' Professor and the founding Director of BEYOND: Center for Fundamental Concepts in Science at Arizona State University (ASU). He is also Principal Investigator in the Center for the Convergence of Physical Science and Cancer Biology and co-director of ASU's Cosmology Initiative. His research has spanned the fields of cosmology, gravitation, quantum field theory, astrobiology, and cancer research, with particular emphasis on black holes, the origin of the universe, the origin of life, and the origin of cancer – topics on which he has authored or co-authored 30 books. He is a Member of the Order of Australia, and the recipient of the Templeton Prize, the Bicentenary Medal of Chile, the Robinson Cosmology Prize, the Faraday Prize of The Royal Society, and the Kelvin Medal of the UK Institute of Physics. The asteroid 1992 OG was officially named "(6870) Pauldavies" in his honor.

MARCELO GLEISER is Appleton Professor of Natural Philosophy and professor of physics and astronomy at Dartmouth College. His research interests include the physics of the early universe, the properties of solitons in classical and quantum field theories, and questions related to the origins of life and self-organizing complexity. He is a fellow of the American Physical Society and an elected member of the Brazilian Academy of Philosophy. He serves on the editorial board of *National Geographic* magazine. His two science series for Brazil's TV Globo were watched by more than 30 million viewers. He writes a weekly science column for a Brazilian newspaper and is the co-founder of a science and culture blog hosted by National Public Radio.

STUART A. KAUFFMAN, a biologist who was trained as a medical doctor, is Finland Distinguished Professor at Tampere University of Technology. He holds joint appointments as a visiting distinguished research professor at the University of Vermont in the College of Medicine and the College of Mathematical and Engineering Sciences. He is a fellow of the Royal Society of Canada, was awarded an honorary degree by the Catholic University of Louvain, was a MacArthur Fellow from 1987 to 1992, and received the Gold Medal of the Academia Lincea Rome. The former co-editor-in-chief of the *Journal of Theoretical Biology*, he has served on the editorial boards of many other journals and has written four books. His founding patents about what is sometimes called "molecular diversity" helped spawn a field known as combinatorial chemistry. He is well known for work on self-organization in evolution, complexity theory, and collectively autocatalytic sets for the origin of life. Recent work with G. Longo and M. Montevil (see Chapter 8) suggests that no laws entail the evolution of the biosphere.

DAVID C. KRAKAUER is Professor of Genetics at the University of Wisconsin-Madison, Director of the Wisconsin Institute for Discovery, and an external professor at the Santa Fe Institute. His research focuses on the evolutionary history of information processing mechanisms in adaptive systems. The current emphasis of his work is on robust information transmission and signaling dynamics, particularly their role in constructing novel, higher level structures, such as social systems and language. He moved on to the Santa Fe Institute as a professor in 2002 and was made faculty chair in 2009. He is a member of the editorial boards of the *Journal of Theoretical Biology, Theory in Biosciences, Biology*

Digest, Interdisciplinary Science Review, Monographs in Mathematical Biology and *Primers in Complex Systems*.

CHARLES H. LINEWEAVER is an associate professor at the Australian National University's Planetary Science Institute (PSI), a joint venture of the ANU Research School of Astronomy and Astrophysics and Research School of Earth Science. His research involves analysis of the statistical distribution of exoplanets, the cosmic microwave background radiation, and cosmological prerequisites for the formation of terrestrial planets and life. He is a member of the editorial board of *Astrobiology Magazine*.

SETH LLOYD, Professor of Mechanical Engineering and Engineering Systems at the Massachusetts Institute of Technology, is interested in the role information plays in physical systems, particularly systems in the quantum realm. He is a principal investigator at the Research Laboratory of Electronics in Cambridge, Massachusetts, and an adjunct professor at the Santa Fe Institute. His pioneering research in the fields of quantum computation and quantum communications resulted in the first technologically feasible design for a quantum computer, and he also has demonstrated the viability of quantum analog computation, proven quantum analogs of Claude Shannon's noisy channel theorem, and designed novel methods for quantum error correction and noise reduction. He is a fellow of the American Physical Society.

MICHAEL RUSE is a philosopher of science who has found in evolution a kind of *Weltanschauung*, a world picture that gives meaning to life. He is one of the foremost contemporary Darwin scholars. Ruse currently teaches at Florida State University. He has honorary degrees from the University of Bergen in Norway and McMaster University, and is a fellow of both the Royal Society of Canada and the American Association for the Advancement of Science (AAAS). The founding editor of *Biology and Philosophy*, he serves on the editorial boards of eight other journals as well as serving as general editor of the Cambridge Studies in Philosophy and Biology. He is a former president of the History and Philosophy of Science Section of the AAAS and presently an associate of the Zygon Center for Religion and Science at the Lutheran School of Theology in Chicago.

D. ERIC SMITH is an external professor at the Santa Fe Institute. He works on problems of self-organization in thermal, chemical, and biological systems. His particular focus at present is the statistical mechanics of the transition from the geochemistry of the early Earth to the first levels of biological organization with an emphasis on the emergence of the metabolic network. D. E. Smith worked in physical, non-linear, and statistical acoustics for seven years at the university's Applied Research Laboratories and at the Los Alamos National Laboratory before joining the Santa Fe faculty in 2000.

WILLIAM C. WIMSATT is emeritus Professor of Philosophy and in the Committee on Evolutionary Biology at the University of Chicago, and is Winton Chair in the Center for Philosophy of Science at the University of Minnesota. He is known for his groundbreaking work in the philosophy of biology. His research and writing center on a cluster of problems arising in the analysis of the structure, behavior, and evolution of complex, functionally organized systems. A fellow of the American Association for the Advancement of Science, he is the recipient of awards for curriculum innovation and outstanding graduate teaching. He serves on the scientific advisory board of the Land Institute, the board of advisors of the Philosophical Gourmet (Leiter) Report, as advisory editor to *Biological Theory*, and on the editorial boards of *Foundations of Science, Journal of Cognition and Culture*, the electronic *Evolutionary Psychology*, and of the BioQUEST Educational Consortium.

DAVID WOLPERT is currently in the Information Sciences Division of Los Alamos National Laboratory and an external professor at the Santa Fe Institute. Previously he was Ulam Scholar at the Center for Non-linear Studies in Los Alamos. Earlier in his career, he was at the NASA Ames Research Center and a consulting professor at Stanford University, where he formed the Collective Intelligence group. He has worked at IBM and at a data mining startup, and been an external faculty member at numerous institutions. His current research focuses on game theory, the application of machine learning to both optimization and Monte Carlo methods, complexity measures, modeling evolution of technology, information theory, and the foundations of physics and inference. He is the author of three books, three patents, and more than one hundred refereed papers, and has won numerous awards.

Part I Introduction

I What is complexity? Is it increasing?

C. H. Lineweaver, P. C. W. Davies, and M. Ruse

One of the principal objects of theoretical research is to find the point of view from which the subject appears in the greatest simplicity.

(Gibbs, 1881)

Most people don't need to be persuaded that the physical world is bewilderingly complex. Everywhere we look, from molecules to clusters of galaxies, we see layer upon layer of complex structures and complex processes. The success of the scientific enterprise over the last 300 years largely stems from the assumption that beneath the surface complexity of the universe lies an elegant mathematical simplicity. The spectacular progress in particle and atomic physics, for example, comes from neglecting the complexity of materials and focusing on their relatively simple components. Similarly, the amazing advances in cosmology mostly ignore the complications of galactic structure and treat the universe in a simplified, averaged-out, approximation. Such simplified treatments, though they have carried us far, sooner or later confront the stark reality that many everyday phenomena are formidably complex and cannot be captured by traditional reductionist approaches. The most obvious and important example is biology. The techniques of particle physics or cosmology fail utterly to describe the nature and origin of biological complexity. Darwinian evolution gives us an understanding of how biological complexity arose, but is less capable of providing a general principle of why it arises. "Survival of the fittest" is not necessarily "survival of the most complex".

Complexity and the Arrow of Time, ed. Charles H. Lineweaver, Paul C. W. Davies and Michael Ruse. Published by Cambridge University Press.
© Cambridge University Press 2013.

In recent decades, partly due to the ready availability of fast and powerful computation, scientists have increasingly begun to seek general principles governing complexity. Although this program remains a work in progress, some deep issues of principle have emerged. How are the different forms of complexity – for example, the chaotic jumble of a pile of rocks versus the exquisite organization of a living organism – related? Secondly, does complexity have a general tendency to increase over time, and if so, when? Cosmologists agree that one second after the big bang the universe consisted of a simple soup of subatomic particles bathed in uniform radiation, raising the question of how the many levels of complexity originated that emerged over time as the universe evolved. In biological evolution too, complexity seems to rise and rise, and yet this trend is notoriously hard to pin down. If there is a complexity "arrow of time" to place alongside the thermodynamic and cosmological arrows of time, it has yet to be elucidated precisely.

The contributors to this volume tackle the foregoing issues head-on, and grapple with the questions: "What is complexity?" and "Is it increasing?". In the tradition of Lord Kelvin, the physical scientists tend toward the view that without a quantitative definition of complexity we won't be able to measure it precisely and *a fortiori* won't know if it is increasing:

> ... *when you can measure what you are speaking about, and express it in numbers, you know something about it; but when you cannot measure it, when you cannot express it in numbers, your knowledge is of a meagre and unsatisfactory kind; it may be the beginning of knowledge, but you have scarcely in your thoughts advanced to the stage of science, whatever the matter may be.*
>
> (Kelvin, 1883)

In the first section of the book, the physicists attempt to construct a definition of complexity – one that can be measured and is transdisciplinary. They want to provide a recipe for measuring complexity while keeping their urges for speculative interpretations on a short leash. The problem, however, is that there are many complexity-generating processes and many forms of complexity (e.g. see Clayton, Chapter 14). Without a unified definition that permits application across a range of disciplines, speculative short leashes are moot. Biologists are less troubled by the lack of a unified definition. After all, shouldn't we expect complexity to be complex? As Conway Morris points out, "First there were bacteria, now there is New York". Even without a rigorous definition, or a broadly acceptable way to measure it, isn't it *qualitatively* obvious that biological complexity has increased? Do we really need to wait for a precise definition to think about complexity and its limits?

Although advances in the study and simulation of complex systems have thrown the problems into sharp relief, Ruse's historical review reminds us that wrestling with complexity is not a new sport. He describes how evolutionary biologists from Darwin to Dawkins have tried to sharpen the intuitive notion of biological complexity. Darwin repeatedly picked at the notion like one picks at a scab. Subsequently, five generations of the world's finest evolutionary biologists have failed to agree on a precise metric to quantify biological complexity. A few plausible proxies have been explored. One is genome length. However, scientists predisposed to believe that humans are the culmination of the animal kingdom (as Darwin believed) are not favorably inclined towards this proxy since (as Davies points out): "*Salamanders have genomes much larger than humans*". The lack of a tight correlation between genome length and our subjective assessment of phenotypic complexity is known as the C-value enigma (Gregory, 2001). Wimsatt describes a specific kind of genomic complexity that he calls "generative entrenchment". It is the genomic complexity of multicellular organisms that has accumulated over hundreds of millions of years and controls ontogeny. Much of this regulatory complexity has been found recently in introns (ENCODE Project Consortium, 2012), but quantifying it is a daunting task.

Comparing the complexities of different organisms can be confusing. For example, Ruse describes a proxy for complexity based on the number of cell types in a multicellular organism. But if this proxy is used, it leaves us unsure of how to quantify the extreme diversity and complexity of unicellular extremophiles. An alternative approach is to focus on ecosystems rather than specific organisms. As Smith (Chapter 9) discusses, the metabolic complexity or the number of species in an ecosystem may be more important than the number of cell types in a single species, in which case a bacterial mat could be judged more complex than a worm or a vertebrate brain.

Both Wolpert and Conway Morris stress the importance of considering the degree of non-uniformity across different scales of size and different parts of a system when assigning a level of complexity. Thus biological complexity can be found in the increased specialization of body parts such as the duplication and subsequent differentiation of animal limbs, in the relationships between species, and in networks of ecosystems. Although the degree of specialization seems a reasonable proxy for biological complexity, as Conway Morris reminds us, there are many examples in the history of life on Earth in which specialization has led to extinction, while simplification has led to adaptive success and survival. Thus, macro-evolution displays trends towards both complexity and simplicity.

The late Stephen J. Gould strongly cautioned against the temptation to ascribe any overall directionality to biological evolution (Gould, 1996). In a famous metaphor he compared the evolution of a species in complexity space to the random walk of a drunkard who starts out leaning against a wall, but whose movement towards the gutter is unrestricted. Gould remarked that in biology there is a "wall" of minimal complexity necessary for life to function at all. He then argued that a mere constraining wall of minimal complexity does not amount to a driving force towards increasing complexity. If there is any sort of driving force at work in evolution it might be more accurately described as a tendency towards diversity, or an entropically-driven tendency to spread out in possibility space, occupying new niches made available by a varying environment.

A similar position is adopted by McShea and Brandon in their 2010 book: *Biology's First Law,* They describe a *"zero-force evolutionary law"* in which *"diversity and complexity arise by the simple accumulation of accidents"*. They argue that when you start with a clean white picket fence, it gets dirty – paint will peel here and there – increasing the "complexity" of the fence. This somewhat impoverished view regards complexity as being based on variation by random drift, and highlights an obvious question: are diversity and complexity synonymous? If our definition of biological complexity refers only to variation without the environmental-information-inserting influence of selection, we have omitted a key factor in biological evolution, which requires both variation and selection. An increase in variation without involving selection merely describes an increase in entropy (a pristine fence naturally degenerating into random defects) and an approach to equilibrium rather than an increase of complexity. This is simply the second law of thermodynamics and fails to capture the key quality of complexity as more than mere randomness. To make progress on this issue, we need to distinguish between variation that brings the system closer to equilibrium and variation that takes it further from equilibrium.

Ruse summarizes McShea and Brandon's views and relates them to Gould's wall of minimal complexity:

> *[McShea and Brandon] say that complexity is "a function only of the amount of differentiation among parts within an individual". Elsewhere they say ""complexity" just means number of parts types or degree of differentiation among parts". They are very careful to specify that this has nothing to do with adaptation. Indeed they say "in our usage, even functionless, useless, part types contribute to complexity. Even maladaptive differentiation is pure complexity". How could this complexity come about? It*

> *all seems to be a matter of randomness. With Gould, and I think*
> *with Spencer, they simply believe that over time more and more*
> *things will happen and pieces will be produced and thus complex-*
> *ity will emerge. It is the inevitability of the drunkard falling into*
> *the gutter.*

What is clear from these exchanges is that one cannot isolate the complexity of biological organisms from the complexity of their environment. Whether the inhabitants of newly available niches will become simpler or more complex depends on the complexity and simplicity of those new niches. Random walks that elevate complexity are driven by the complexity of the environment being walked in. When environments get more varied and complex, so too do the inhabitants. If the environment gets simpler (closer to equilibrium with fewer gradients) then the inhabitants get simpler as selection pressure is removed and the information in their genomes drifts away.

Part of the problem of quantifying biological complexity is that it seems impossible to put a measure on the number of new niches, let alone the number of all possible niches. How many ways of making a living are there? If the number increases with time, where does this increase come from, and is there any meaningful upper bound?

Kauffman and Conway Morris both address the issue of the potential for biological complexity afforded by an expansion of the niche space. The former opines that

> *we cannot know what the set of all the possibilities of the evolu-*
> *tion of the biosphere are. Not only do we not know what WILL*
> *happen, we don't even know what CAN happen.*

Also, by coining the memorable phrase: *"The uses of a screw-driver cannot be listed algorithmically"* Kauffman emphasizes the unlimited nature of exaptations and the impossibility of listing them in advance. Conway Morris has a different view. Based on his estimation of the importance and ubiquity of convergence, he claims that the space of phenotypes is not only saturable, in many cases it is already saturated. He contends that if the tape of life were played again, much the same results would ensue. To bolster his claim, Conway Morris presents evidence that *"biological systems are near, and in some cases have clearly reached, the limits of complexity"*. One exception to this – in Conway Morris's view – are human beings, who don't seem to have reached the limits to their complexity.

Kauffman contests Conway Morris's saturation claim. Unfortunately, in the absence of a measure of biological complexity it is hard to decide between these competing points of view. Davies at least sides with Kauffman, when he writes

> *At any given time there is only a finite number of species, but the number of potential species is almost limitless, as Stuart Kauffman explains so well in his chapter. Therefore random mutations are far more likely to discover a new species than to recreate old ones. No mystery there.*

If the proximate origin of the increase of biological complexity lies in the adaptations that take advantage of new niches, the ultimate origin must be in the mechanism that supplies the new niches. What mechanism controls the number of new niches? Obviously changes in the environment play a key role. However, the environment of an organism includes not only the physical variables (temperature, rainfall, etc.) but the existence and activities of other organisms in the ecosystem. As a specific example of how a new niche drives the evolution of new adaptations, see the detailed genomic analysis of *E. coli* evolution in a new citrate-rich, glucose-limited medium (Blount et al., 2012).

Lineweaver argues that one can simplify the analysis of complexity because all the information and complexity in biological organisms ultimately derives from the physical environment. In addition, since cultural information and complexity depends on biological complexity, cultural complexity, too, ultimately depends on physical complexity. If this causal chain is correct, then we can understand the source and nature of complexity by focusing on the evolution of the complexity of the physical universe (galaxies, stars, and planets).

In his chapter, Krakauer gives an explicit description of how natural selection transfers the information from the environment to the genome. In addition, when the environmental information is changing too rapidly to be incorporated into genes (i.e. environmental changes on timescales shorter than a generation) he describes how environmental information can become incorporated into biological information storage facilities with much faster turnover times: i.e. brains. Thus both Lineweaver and Krakauer argue that biological information is not created de novo, but concentrated in organisms via evolutionary and cognitive processes. The ultimate source of biological information and complexity must then be found in the physical universe. Although it is hard to fault this line of reasoning, it carries the implicit assumption that information and complexity are in some sense passed on like a token from environment to organism. The argument shifts the explanation for biological information and complexity to the environment. But this is just passing the buck or moving the bump in the carpet unless we can answer the question of how the information and complexity got into the environment in the first place.

One aspect of the problem has general agreement. Complexity cannot increase in time without a source of free energy to generate (or transfer) it. This is possible only if the universe is not in a state of thermodynamic equilibrium (sometimes referred to as the heat death). Well-known examples of the emergence of complexity and the concomitant rise of entropy (or fall of free energy) to pay for it are (i) the organized structure of a hurricane, which is possible only by the existence (and low entropy) of pressure, temperature, and humidity gradients, and (ii) the origin of life driven by the exploitation of some form of chemical redox potential. Mineral evolution is also an example of increasing abiotic complexity that depends on the flow of free energy (Hazen *et al.*, 2008). The largest variety of minerals is found near ore bodies where there has been high free energy through-put from the condensation of hot fluids driven by a temperature gradient.

These ideas support Chaisson's use of specific free energy flow as a proxy for complexity. Energy rate density correlates with system complexity. As Chaisson points out, this has the advantage that energy (as opposed to complexity) is something we know how to measure.

Because physical complexity requires the exploitation of free energy gradients, the growth in complexity of any kind is linked to free energy decrease and entropy increase, in conformity with the second law of thermodynamics. However, just because the growth of complexity is consistent with the second law does not mean it is explained by the second law. Many authors have recognized that entropy and the second law have some fundamental connection with complexity. However, it is not a simple inverse relationship. Entropy is best thought of as the degree of disorder in a system, implying that it is *order*, rather than complexity, that plays the inverse role. Hence there is no incompatibility between advancing complexity and increasing entropy.

Wolpert illustrates this point explicitly with a simple example. He considers a hollow sphere filled homogeneously with a few large balls and many small ones. Random jiggling then moves the large balls outwards to hug the walls of the sphere, leaving the small balls in the middle. This slightly more organized and complex state leaves the large balls with a lower entropy. However, the entropy of the rest of the system increases because the small balls occupy a greater volume. Since there are so many more small balls, the increase in volume available to them more than compensates for the decrease in volume available to the large balls, in that the total entropy increases – the total disorder increases.

These considerations raise an interesting question. Given that the price of complexity is (as with all physical processes) additional entropy generated in the wider universe, does the growth of complexity accelerate or retard the growth of entropy in the universe – hastening or slowing the heat death? Does the answer depend on the type of complexity involved? Gleiser presents a model in which organized complexity manifested as a specific form of spatiotemporal patterning emerges spontaneously in a non-linear system, and shows that energy equipartition is slowed as a result. He writes that *"the emergence of self-organizing spatiotemporal structures works like an entropy sink: locally, they decrease the system's entropy, while globally they slow its evolution to its final maximum value"*.

But the use of the term "spontaneously" here needs nuancing, because the patterns emerge only because the systems are driven by the out-of-equilibrium, low-entropy initial conditions. Just as in the case of Wolpert's pattern of large balls, an entropic price has been paid for the organization. To call such structures "self-organizing" implicitly includes the out-of-equilibrium initial conditions as part of "self".

Gleiser's structures (solitons and oscillons) which "slow [the evolution of entropy] to its final maximum value" seem to be one example of what Dyson (1971) has called "hangups" in the flow of energy. But whether there are complex structures that slow entropy production is controversial. For example, it is often assumed that all far from equilibrium dissipative structures on Earth (including all life forms) dissipate free energy faster than would be the case without them (Ulanowicz & Hannon, 1987; Meysman & Bruers, 2010; Kleidon, 2012). On this view, without life on Earth, there would be more free energy streaming out from the Sun into interstellar space. The entropy of the universe would be lower and there would be more free energy able to produce complexity elsewhere.

Although most contributors agree that the complexity of the universe has risen since the big bang, could this rise really be just a type of pyramid scheme? If we focus only on the richest people at the top of the pyramid, it seems obvious that the scheme is making people richer. Similarly, we seem to have a natural tendency to focus our attention on the most complex structures – the top of the complexity pyramid. In asking the question, is complexity increasing?, are we mesmerized by the tiniest most complex needle in an overwhelmingly simple haystack? Progress can be made if we divide our original question, is complexity increasing? into two separate questions: is the average complexity of the universe increasing? and is the complexity of the most complex objects or systems increasing?

Without a definition of complexity that can be applied to a relatively simple environment that is approaching equilibrium, the answer to the first question is unclear. Whether or not the most complex objects will continue indefinitely to get more complex depends on the entropy gap ΔS between the actual entropy of the universe and the maximum entropy (Lineweaver). This gap provides the free energy to keep the complexity of the most complex objects increasing. Chaisson argues that expansion itself causes free energy to flow and complexity to rise. On this view, complexity will continue to increase as long as the universe expands. Lineweaver takes the position that the universe is approaching an eventual heat death and as it does, the entropy gap approaches zero, placing a bound on the total complexity in the universe. Lloyd points out that the answer depends on whether the universe is dominated by dark energy (as present observations suggest):

> Whether or not free energy will continue to be available within the observable universe for all time is an open physical and cosmological question. If the observed cosmological term that is causing the accelerating expansion of the universe remains bounded away from zero, then the answer is apparently No.
>
> (Lloyd)

Davies makes the same point that the far future of the universe will be one of increasing simplicity as a result of the accelerating expansion. He frames this prediction in the context of gravitational entropy, a set of ideas dating to the work of Hawking and Penrose in the 1970s, and still unresolved.

> *The story that we take away from this modern cosmological view*
> *is that the universe started out bland and featureless, and over*
> *time has evolved increasing richness of structure and process, and*
> *ever greater complexity of systems, on all scales of size except the*
> *largest, where uniformity still reigns.*
>
> *(Davies)*

On the other hand, the systems of ever greater complexity occupy a tiny fraction ($< 10^{-30}$?) of the volume of the universe. If we are really only interested in the tiniest most complex needle in an overwhelmingly simple haystack then we are missing the big picture unless we can justify that the emergence, maintenance, and increase of complexity of a small part of the universe is completely decoupled from the rest of the universe. However, complexification of tiny regions is not decoupled from the rest of the universe since producing and maintaining local complexity requires consumption of free energy, local entropy decrease, global entropy increase, and, with that, a decreased ability to produce complexity elsewhere.

One big impediment to using entropic considerations to explore complexity is the lack of a precise formula for gravitational entropy – whose low initial value seems to be at the origin of all free energy flows needed to produce complexity.

> *any attempt to identify gravitational complexity with nega-*
> *tive gravitational entropy runs into definitional problems from*
> *both ends – neither complexity nor gravitational entropy has*
> *universally-accepted general-purpose definitions at this time.*
>
> *(Davies)*

Lloyd sees the universe as a quantum computer and

> *as long as our universe is supplied with a suitable source of ini-*
> *tial random information as program, it must necessarily produce*
> *complex structures with non-zero probability.*
>
> *(Lloyd)*

Several authors agreed that human brains are the most complex entities in the universe – we are at the top of the pyramid. Under this definition, entropy increase can provide meaning to humans. However, in our attempts to understand complexity objectively, we should take a humble approach and be wary of how self-serving such a notion of complexity is. There is also a potential theological bias, as Clayton points out:

> one could intuitively feel that a universe of increasing complexity is the sort of universe one would expect if religious views of ultimate reality are correct.
>
> *(Clayton)*

Arising from a symposium, "Is There a General Principle of Increasing Complexity?", held at the BEYOND Center for Fundamental Concepts in Science at Arizona State University in December 2010 and sponsored by the John Templeton Foundation, this volume pulls our thoughts in two opposite directions. The physicists describe attempts to find a unified definition of complexity, while the biologists and complexity scientists describe how complex and non-unified complexity is. Unable to bridge this gap, complexity can seem like a delusion in pursuit of a theory. Perhaps, like beauty, the subjectivity of complexity will keep it immune from any single objective scientific assessment. We have not resolved the questions: What is complexity? Is it increasing? Does it have an ultimate limit? – but we have made our confusion about these questions more explicit.

REFERENCES

Blount, Z. D., Barrick, J. E., Davidson, C. J., & Lenski, R. E. (2012). Genomic analysis of a key innovation in an experimental *Escherichia coli* population. *Nature*, **489**, 513–518, doi:10.1038/nature11514.

Dyson, F. J. (1971). Energy in the Universe. *Scientific American*, **225**, Issue 3, pp. 50–59.

ENCODE Project Consortium, (2012). An integrated encyclopedia of DNA elements in the human genome. *Nature*, **489**, 57–74.

Gibbs, J. (1881). From Gibbs's letter accepting the Rumford Medal (1881). Quoted in A. L. Mackay (1994). *Dictionary of Scientific Quotations* (London: Taylor & Francis).

Gregory, T. R. (2001). Coincidence, coevolution or causation? DNA content, cell size, and the C-value enigma. *Biological Review*, **76**, 65–101.

Gould, S. J. (1996). *Full House: The Spread of Excellence from Plato to Darwin.* New York: Harmony Books.

Hazen, R. M., Papineau, D., Bleeker, W. *et al.* (2008). Mineral Evolution. *Amer. Mineralogist*, **93**, 1693–1720.

Kelvin, W. T. (1883). The Six Gateways of Knowledge, Presidential Address to the Birmingham and Midland Institute, Birmingham (3 Oct 1883). In *Popular Lectures and Addresses* (1891), **1**, 280.

Kleidon, A. (2012). How does the Earth system generate and maintain thermodynamic disequilibrium and what does it imply for the future of the planet? *Phil. Trans. R. Soc A*, **370**, 1012–1040.

McShea, D. W. & Brandon, R. N. (2010). *Biology's First Law.* Chicago: University of Chicago Press.

Meysman, F. J. R. & Bruers, S. (2010). Ecosystem functioning and maximum entropy production: a quantitative test of hypotheses, *Phil. Trans. R. Soc. B.* **365**, 1405–1416.

Ulanowicz, R. E. & Hannon, B. M. (1987). Life and the production of entropy. *Proc. Royal Soc. London. Series B, Biological Sciences*, **232**, No. 1267, 181–192.

Part II   Cosmological and physical perspectives

2 Directionality principles from cancer to cosmology

P. C. W. Davies

2.1 THE BIG PICTURE

One of the gloomiest scientific predictions of all time was made in 1852 by the physicist William Thompson (later Lord Kelvin) (Thompson, 1852). From a consideration of the laws of thermodynamics, and the nature of entropy, Thompson declared that the universe is dying. The second law of thermodynamics, which had been formulated a few years earlier by Clausius, Maxwell, Boltzmann and others (see, for example, Atkins, 2010), states that in a closed physical system, the total entropy – roughly a measure of disorder – can never decrease. All physical processes, while they may produce a fall of entropy in a local region, always entail a rise of entropy somewhere else to pay for it, so that when the account is tallied, the total entropy will be seen to have risen. Applied to the universe as a whole, the second law predicts an inexorable rise of the overall entropy with time, and a concomitant growth in disorder. The one-way slide of the universe towards total disorder – popularly known as the heat death of the universe – imprints upon it an irreversible arrow of time. One need look no further than the Sun, slowly burning through its stock of nuclear fuel, radiating heat and light irreversibly into the cold depths of space, to see an infinitesimal contribution to the approaching heat death. Eventually its fuel will be exhausted, and the Sun will die, along with all other stars when their time has come. The sense of futility and pointlessness that the dying universe scenario engenders in some commentators was eloquently captured by Bertrand Russell

Complexity and the Arrow of Time, ed. Charles H. Lineweaver, Paul C. W. Davies and Michael Ruse. Published by Cambridge University Press.
© Cambridge University Press 2013.

in his book *Why I Am Not a Christian* (1957) and has been echoed in recent years in the writings of Peter Atkins (1986).

In Thompson's day, cosmology was not a well formulated science. In particular, the universe was assumed to be static, and the discussion of the cosmological heat death prediction must be understood in that context. Today we know that the universe is expanding, and sprang into existence (at least in its present form) with a big bang about 13.8 billion years ago. The rise of entropy must now be defined in reference to an expanding volume of space. A simple view of the second law of thermodynamics is that the universe began in a low-entropy state, and the entropy has been rising ever since, and will go on rising into the future, although whether or not the universe ever attains a maximum entropy state of thermodynamic equilibrium is a subtle one that I shall return to later.

What do we know about the state of the universe near the beginning? The cosmic microwave background radiation (CMB) – the famous fading afterglow of the big bang – provides a snapshot of what the universe was like, about a mere 380 000 years after the big bang. The spectrum of the CMB is textbook Planckian, that is, it is characteristic of a state of matter and radiation in thermodynamic equilibrium. This seems wrong: the universe is supposed to have started out in a low-entropy state, yet thermodynamic equilibrium is a maximum entropy state (Davies, 1975). How has the universe gone from maximum entropy at the start to less-than-maximum entropy today – the wrong way according to the second law of thermodynamics?

2.2 GRAVITATIONAL ENTROPY AND GRAVITATIONAL COMPLEXITY

The answer to the question of how primordial thermodynamic equilibrium has given way to disequilibrium over time has been known for several decades, and it concerns the curious nature of gravitation. To be sure, matter and radiation may have been in thermodynamic equilibrium at the outset, but the gravitational field of the universe was in fact very far from such a state. The CMB, plus astronomical observations of the distribution of galaxies, show that the cosmological material is distributed remarkably uniformly through space, and the gravitational field of this material is likewise almost exactly uniform. A uniform gravitational field, being relatively ordered, represents a low-entropy state (Penrose, 1979). In a self-gravitating system, matter that is free to do so will grow increasingly clumpy with time. ("Free to do so" implies mechanisms to divest the matter of heat radiation and angular momentum, which serve to permit gravitational contraction.) Thus our Galaxy formed from a diffuse nebula, while the Solar System formed by the contraction of a cloud of gas and dust. One can discern the faint beginnings of this growth of clumpiness at 380 000 years, from slight irregularities in the CMB. Over time these irregularities have become amplified by gravitation, producing structure on all scales of size from cosmological down to galactic. The ultimate end state of this gravitational aggregation is shrinkage into a black hole. In 1975, Stephen Hawking provided a precise formula for the entropy of a black hole (Hawking, 1975), widely believed to be the maximum possible entropy of a gravitating system, and confirming that the entropy of the more diffuse distribution of matter that characterized the early universe was much less than the maximum (Penrose, 1979). Hawking's formula, which builds on earlier work by Jacob Bekenstein (1973), identifies the *area* of the event horizon of the black hole with its entropy. (The event horizon of a black hole is, roughly speaking, its surface. Events that occur inside the horizon cannot be observed from outside.) Thus if two black holes collide and merge, the total event horizon area afterwards is never less than the sum of the event horizons of the individual black holes before the collision, in perfect accord with the second law of thermodynamics (Hawking, 1971).

Although a black hole may represent the maximum gravitational entropy, it is not necessarily a state of maximum total entropy. In the same work that Hawking derived an explicit formula for the entropy of a black hole, he also demonstrated that a black hole has an intrinsic temperature and can therefore emit heat radiation, lose energy, and shrink. If a black hole is free to evaporate away completely into low-mass particles (e.g. photons) via the Hawking process, then the entropy of the emitted radiation is somewhat greater than that of the original black hole (Davies, 1977).

The puzzle of the second law applied to the universe is therefore solved. Matter and radiation might have started out in a maximum entropy state, but the gravitational field, being relatively smooth, was in a low-entropy state. Added together, the gravitational, matter and radiation entropies have always been rising (Davies, 1975; Lineweaver and Egan, 2008; Egan and Lineweaver, 2010). The departure of the latter two components from equilibrium since 380 000 years may be traced, more or less, to the clumping effects of gravitation. For example, the shrinkage of gas clouds heats the material, creating the type of thermal gradients and heat flow familiar from the Sun radiating into cold space. The thermodynamic disequilibrium which this represents is paid for by the growing disorder of the gravitational field, which becomes more and more inhomogeneous, representing a rise in gravitational entropy.

All these things have been intensively studied, puzzled over and discussed for some decades (see, for example, Davies, 1975; Penrose, 1979), and I summarize them here merely to provide a context for the real puzzle: the evolution of complexity. The early universe, as I have explained, consisted of gas and radiation at a common temperature distributed extremely smoothly throughout space. All the structure we now observe – the galaxies, stars, planets, and moons in their teeming diversity – have emerged since the hot, simple, primeval phase. The story that we take away from this modern cosmological view is that the universe started out bland and featureless, and over time has evolved increasing richness of structure and process, and ever greater complexity of systems, on all scales of size except the largest, where uniformity still reigns. (If by largest one refers to about 100 Mpc and above. On the other hand, a god's-eye-view of all of reality might reveal a multiverse of exceedingly heterogeneous "bubble" universes.)

There are many physical processes that contribute locally to this overall trend, but the fountainhead that drives them all may ultimately be traced back to gravitation – the smooth order of the original state, and the tendency of gravitation to shrink and clumpify matter, and so release energy able to drive all the aforementioned complexifying processes.

We thus have two competing narratives for the great cosmic story. One speaks of inexorable degeneration and decay, and descent from order into chaos at the behest of the second law of thermodynamics, the other of growing complexification and enrichment. Are these accounts in conflict? The answer is no. As I have explained, the complexification of the universe is paid for by a rise in the entropy of the gravitational field, so whereas matter and radiation enjoy sustained "free energy" (roughly speaking, energy available to do work) by virtue of the gravitational field to drive complicated processes, the gravitational field itself pays the price in its being disordered. So the total entropy of the universe rises even as the richness, complexity, and diversity of its contents goes up.

We now hit the central point I want to make about complexity. A low-entropy gravitational field is simple in form, whereas a high-entropy state is complex. In the case of matter in which the effects of gravitation may be ignored, it is usually (but not always) the other way around. Thus a box of gas in thermodynamic equilibrium (maximum entropy) is in a simple, uniform state on a macroscopic scale. (It is maximally disordered, or chaotic, on a microscopic scale, however.) If gas has large-scale (coarse-grained) structural complexity, then to the extent that gravitation can be ignored, it is in a less-than-maximum entropy state. This "back-to-front" nature of gravitation is also manifested in the fact that the specific heat of self-gravitating systems is negative. For example, if a star loses energy, it gets hotter, not cooler. The same thing happens when a black hole radiates and shrinks: its temperature rises (Hawking, 1975).

Unfortunately this simple picture of gravitational clumping comes with a corollary, and it is an important one. While ordinary matter and radiation energy exert a gravitational pull, driving the clumping process, there are known to be forms of energy that create a gravitational push, i.e. repulsion. Gravitational repulsion has the effect of making the universe on a large scale grow *smoother*, not clumpier, with time. Until recently it was possible to ignore this complication, because the universe seemed to be dominated by positively gravitating stuff. Now it is fairly certain that it is the other way about. For roughly half its history the universe was indeed dominated by gravitational attraction on the largest scale of size, with the consequences for the growth of complexity that I have explained. Now we know that a few billion years ago that state of affairs gave way to an overall cosmic repulsion, causing the expansion of the universe to speed up. The source of this cosmic repulsion is something called *dark energy*, which evidently fills empty space; this is not the place to discuss its origin or properties beyond remarking that although it dominates on the largest cosmological scale, in local regions (e.g. within our Galaxy) its effect is negligible, and the gravitational clumping story runs much as before. But in regard to the ultimate fate of the universe, it now looks as if it will in fact be returned to a state of featureless uniformity, indeed, almost complete emptiness, but expanding at a much slower rate than the early universe. It really is a case of the universe "starting with a bang and ending with a whimper".

As regards the entropy of the gravitational field, once dark energy comes into the picture, it is no longer true that the black hole is the maximum gravitational entropy state. Rather this accolade is reserved for de Sitter space – an exponentially expanding empty space to which it looks as if our universe will tend in the coming few billion years. By analogy with the area of a black hole being a measure of its entropy, so too is the area of the cosmological event horizon accepted as a measure of the gravitational entropy of the universe in its final de Sitter state, and in fact it may be proved that the de Sitter entropy thus defined sets an upper bound to which the total entropy of the universe asymptotes over time (Bousso, 2002). The fact that black holes and cosmological event horizons possess properties analogous to entropy has led to the formulation of a generalized second law of thermodynamics, whereby the total entropy – matter and radiation on the one hand and event horizon area on the other – should never decrease with time (Davies and Davis, 2003; Davies, Davis and Lineweaver, 2003).

It turns out that this story is further complicated if the density of dark energy is a function of time. For example, in the case that it increases with time, the universe expands super-exponentially, and can terminate in a "big rip" singularity, where the expansion rate diverges (Davies, 1988; Caldwell et al., 2003). In the approach to the final singularity, the total cosmological horizon area decreases, and there seems to be no concomitant increase in any other known entropy, gravitational or otherwise, to prevent a steady decrease in total entropy within a given horizon volume, in clear violation of the generalized second law of thermodynamics. If the density of dark energy decreases with time, the universe can eventually reach a state of maximum distension and then begin to shrink, terminating with a big crunch singularity. In that case, once again, the total entropy can go down near the end (Davies, 1988).

Some things I have glossed over in this tale of seesawing cosmic complexity. The first is that the entropy of the gravitational field has never been satisfactorily defined in the general case. The entropy of black holes and of de Sitter space are indeed well defined – these asymptotic end states are characterized by having event horizons, and the entropy is, as explained, simply their area, in appropriate units. Although the tendency for gravitating systems to irreversibly grow clumpy looks like a close analog of the irreversible rise in entropy, and defines its own arrow of time, physicists have never been quite able to put their figure on precisely what is a suitable measure of clumpiness. Some attention has been given to a proposal by Roger Penrose to use the square of the Weyl tensor as a measure of gravitational entropy (Penrose, 1979), but this has not been universally accepted. Certainly the Weyl tensor is not a monotonically increasing function of time under generic circumstances. Worse, in the black hole and de Sitter cases the entropy depends explicitly on Planck's constant of quantum mechanics, yet collapsing and fragmenting clouds of gas look thoroughly classical. In the absence of a quantum theory of gravity, these two aspects of gravitational entropy cannot be knitted together.

Added to these woes is a vagueness about how to define "complexity" properly anyway – a shortcoming that is not confined to the gravitational field. Thus any attempt to identify gravitational complexity with negative gravitational entropy runs into definitional problems from both ends – neither complexity nor gravitational entropy has universally-accepted general-purpose definitions at this time. So whilst we have an intuitive notion that structure, diversity and complication are driven by, and triggered by, gravitational aggregation (locally), and swept to smoothness by dark energy antigravity (globally), so far this account is little more than words. What is certainly lacking is any sort of rigorous theorem about when and under what circumstances cosmological complexity will rise.

2.3 MATERIAL COMPLEXITY

Quite apart from the gravitational story, the history of matter is also one of increasing complexity. At about one second after the big bang, the universe consisted of a uniform soup of protons, neutrons, electrons, and neutrinos. As the universe expanded and cooled, matter not only aggregated into structures, but began a process of progressive differentiation that continues to this day. The first stage was the formation of helium during the first three minutes, so that the chemical composition of the cosmological material became two-component: H and He. With the formation of galaxies about 400 million years later the first stars were born and heavier elements were added to the H and He, and disseminated into the interstellar regions by supernova explosions. Some of these heavier elements formed into ice crystals and dust, thus releasing the potential for an almost unlimited variety of solid material forms, ranging from tiny grains up to planets. Subsequent generations of stars were accompanied by swirling disks of dust and gas that formed planetary systems of surprising diversity.

With the appearance of planets with solid surfaces, the way lay open for the further enrichment of material forms, through crystallization and the formation of amorphous substances. The explosion in the size of the possibility space for physical forms was astronomical, because the number of ways of combining solid structures is superexponential. One need go no further than the humble snowflake to see that even a population of identical ice crystals can be combined into so many hexagonal filigree patterns that no two snowflakes in the history of the Earth are likely to have been identical. A similar story applies to almost all solid structures; no two rocks are the same in their internal composition or external shape, for example. The distribution of solid objects likewise knows no bounds – there is effectively zero probability that the universe contains a planet with rings that match in fine detail those of Saturn, for example. A similar story applies to fluids – no two clouds are the same, no two ocean flow patterns, no two planetary convection patterns (hence no two magnetic field patterns), no two stellar wind patterns, no two cosmic ray showers, . . . The list goes on and on.

The principle that underlies this explosion of diversity can be traced to symmetry breaking. As a general rule, the higher the temperature of a system, the greater is the symmetry. The point is well illustrated by imagining Earth, with its vast array of complex forms, being sealed in a gigantic oven with the thermostat set at 4000 K. As the planet heats up, so the forests catch fire, turning complex life forms into carbon dioxide and featureless ash, the oceans boil, the rocks melt and flow together, losing their distinctive separate identities. Eventually the entire planet liquefies and then vaporizes, so that when the process is complete, the box contains only a uniform gas, or plasma, of various elements. Heat the system to ten billion degrees and even the elements disintegrate and their components recreate the primordial soup of protons, neutrons and electrons. Clearly, a box of uniform gas is much more symmetric – much less structured and differentiated – than the Earth in its present state, with its myriad diverse forms and processes on all scales of size from micro-crystals up to continents, almost all distributed in a higgledy-piggledy manner.

The reverse process – the breaking of symmetries as the temperature is lowered – is an established principle in physics and well understood in terms of the laws of thermodynamics. The breaks occur at distinct *phase transitions*, such as the freezing of water (uniform) to ice (non-uniform, because it defines crystals planes in specific directions). Many such phase transitions are known. As a rule, the lower-temperature phases have less symmetry and so are more complex. Because the universe started out exceedingly hot after the big bang, and cooled as it expanded, the succession of phase transitions and symmetry breaks led to a progressive complexification and differentiation of matter, and an explosive growth in the diversity of physical forms, structures and systems.

This raises the interesting question of whether, and for how long, this process may continue. The further cooling of the universe is unlikely to trigger additional significant symmetry breaks (over and above those already understood by low-temperature physicists, that is), but so long as the stars burn, new material forms and structures will be generated. In the very far future, however, the stars will exhaust their nuclear fuel, and the overall complexity of matter will start to decline. The reason for the decline is manifold. Some stars will form black holes – the simplest objects in the universe – that will swallow up other objects, growing larger as a result. Eventually some large fraction of the matter in the universe will end up inside black holes, effectively eradicated as far as the rest of the universe is concerned. On an immense timescale, even the elementary particles left in the remaining diffuse clouds of gas, or burned-out stars and planets, will decay, and their products will annihilate to form photons. For example, protons are intrinsically unstable, and although their half life is known to exceed 10^{32} years, quantum gravity effects ensure that their half life will not exceed 10^{200} years (Hawking, 1978). Eventually they will all decay into positrons.

The details of the far future of the universe have been much studied (see, for example, Davies, 1994; Adams and Laughlin, 2000).

The results depend somewhat on assumptions about how the universe expands and the nature of the fundamental forces, but all models agree that if the universe goes on expanding forever then a time will come when almost all matter ceases to exist. Even the giant black holes will evaporate via the Hawking process, leaving a universe of dark emptiness, populated only by photons, neutrinos, and gravitons of an ever-diminishing density. Thus the story for matter is similar to that for gravitation: complexity rises sharply as the universe expands and cools – on a timescale of, say, billions of years – but then declines again over a very much longer timescale – between 10^{11} and 10^{65} years (or even longer for black hole evaporation), depending on the process.

2.4 BIOLOGICAL COMPLEXITY

With the formation of planets, the way lay open for the formation of life and the growth of biological complexity. Here we enter controversial territory. Everybody agrees that the biosphere today is far more complex than it was when life began on Earth (nobody knows when that was, but there is general agreement that life had established itself by about 3.5 billion years ago). The increase applies both to the complexity of the most complex life form and to the overall complexity of the biosphere. Thus a human is more complex than a bacterium and a rain forest is more complex than a colony of bacteria. Darwin adopted the metaphor of the tree to describe evolution, with its many branches representing speciation. An evolutionary tree is manifestly time-asymmetric – it looks completely different upside down. Thus mutations may cause one species to split into two by genomic divergence, but we never find two species merging into one (except in the limited sense of endosymbiosis), because the probability of different genome sequences that represent two separate species undergoing just the right mutations to become identical is infinitesimal. But these simple statements conceal many subtleties.

Obviously there can be no absolute law of biological complexity rise: if the Sun were to explode tomorrow and destroy all life, then biological complexity would be re-set to zero! As far as the biosphere as a whole is concerned, the evolution of complexity is by no means a one-way street. Several times the planet has suffered mass extinctions, perhaps due to exogenous shocks from comet impacts, super-volcanoes, nearby supernovae, solar variability, etc., or due to the non-linear dynamics of ecosystems themselves. These episodes have occasionally destroyed 90 per cent or more of species. By some measure of our biosphere's complexity, then, it has generally exhibited an upward trend, but punctuated by sudden setbacks. The tree of life always "points upwards", although the upper branches sometimes receive a severe pruning. The directionality of the tree is not hard to explain. At any given time there is only a finite number of species, but the number of *potential* species is almost limitless, as Stuart Kauffman explains so well in Chapter 8. Therefore random mutations are far more likely to discover a new species than to recreate old ones. No mystery there. If we take as a measure of complexity change, not successive instantaneous sums over all extant species, but the directionality of branching, then there is clearly, for unexceptional reasons, a growth of complexity with time.

When it comes to the complexity of individual organisms, the situation is not so clear-cut. Take any given complex organism today; its lineage will generally, though not always, include simpler precursors. Thus a long-ago human ancestor was a fish. Its long-ago ancestor was a microbe, and so on. But this undeniable fact does not necessarily imply that simple organisms *inevitably* give rise to more complex organisms. In some cases, natural selection favors a return to phenotypic simplicity. For example, fish that live in dark caves have lost the power of sight; they don't need it, so eyes become just one more survival headache – better to be rid of them. Mitochondria were once free-living bacteria, but today they have been stripped down to their bare essentials by evolution and could not survive outside of a eukaryotic cell. The same is true of most parasites – they lose traits that would ensure their autonomous survival because the host can supply what's missing. Parasites "travel light", care of their hosts, and so have become simpler over evolutionary time.

The complexity of organisms may be defined either phenotypically – in terms of their organs – or genotypically – in terms of their genomes. Genomic complexity may be defined in many ways. One measure is the size of the genome. A bacterium has about a million base pairs in its DNA, whereas a human has about a billion. But this crude measure can be misleading. Salamanders have genomes much larger than humans, for example. Also, DNA may possess many noncoding ("junk") regions, and it is far from clear how much of that should be included in assessing genomic complexity. Furthermore, numbers of DNA base pairs per se is a poor measure of complexity. A long segment of DNA consisting of identical base pairs or a short repeating sequence is obviously simpler than a short random sequence. Algorithmic complexity theory provides a framework to take care of this difficulty (Chaitin, 1987), but opinions still differ about what constitutes a suitable measure. Other contenders abound: for example, Lloyd and Pagels (1988) proposed a quantity known as thermodynamic depth, while Bennett (1988) has suggested something called logical depth. I shall not attempt to summarize these distinctions here. Suffice it to say that without an agreed quantitative definition of complexity, it would not be possible to identify with precision when it might increase over time.

Let me turn to the experimental data. Whether we use phenotypic or genotypic measures, the issue before us is whether there is any systematic trend or directionality in the growth of biological complexity over time, and if so, under what circumstances. In Victorian times it was fashionable to talk about evolution as a "ladder of progress", with human beings at the pinnacle. But this may be an illusion founded on chauvinistic expectations of human significance. It is no surprise that if one considers an extant complex organism, such as a human, and looks back along its lineage, its precursors are likely to be less complex. But this does not mean that evolution is *directed towards* complexity, still less towards specific organisms, as has been stressed by Steven Jay Gould (1996). In his analogy with a drunk leaning against a wall, Gould invites us to imagine the drunk staggering along the sidewalk at random, eventually landing in the gutter, not because he is aiming for the gutter, but because the wall bars his movement in one direction and so serves to enhance the chance that he will stray in the other direction. In the case of life, it necessarily starts out simple, so there will be a tendency to accumulate complexity, not for reasons of directionality, but because evolution conducts a random walk through the possibility space which is bounded by a "wall" of minimal complexity. The wall exists for the elementary reason that there is a lower limit, but no obvious upper limit, to the complexity of a living organism.

Set against Gould's argument, however, is the claim that the fossil record displays distinct evolutionary trends towards greater complexity. One of these concerns the encephalization quotient, a measure of the braininess of animals relative to their body size. It does seem that at least in hominid evolution there has been an accelerating trend towards bigger brains and, by assumption, higher intelligence (Bruner et al., 2003). Trends of this nature are perhaps no surprise. Evolution often entails "arms races" that serve to amplify a trait once it appears, thus producing a string of directed evolutionary changes. In addition, the phenomenon of evolutionary convergence will often result in the "discovery" of similar solutions to similar problems via different genotypic pathways, again giving the impression of directional change. It is an open question whether we must simply deal with each case of such a complexity trend on an ad hoc basis, or whether there is a deeper principle of complexification at work that guarantees (under certain well-defined circumstances) an advance in complexity over and above a random walk through the possibility space (Nitecki, 1989). If there is such a principle, then not only will the most complex representative of the biosphere tend to become more complex over time, but the median of the complexity distribution will also shift steadily towards higher values over time. If there is no systematic trend, that does not preclude specific lineages becoming more complex, or new species with still greater complexity emerging, but this random diffusion in the possibility space must not be confused with an overall *directionality* in the degree of complexity.

2.5 THE ORIGIN OF LIFE AND LIFE AS A COSMIC PHENOMENON

Left out of the foregoing discussion is any account of how life got started in the first place. Darwin himself refused to be drawn into this debate. "One might as well speculate about the origin of matter", he once quipped (Darwin, 1863). A century and a half later we remain almost completely ignorant of how a mixture of molecules can become transformed into a living thing (Davies, 2003). In the absence of a definition of life, it is hard to even know what a successful explanation for life's origin would entail. However, one can at least say on general grounds that the pathway from non-life to life must involve a rise in complexity. It is clear that biological complexity is distinctively different from other forms of complexity, for example, that of a chaotic system or a Boltzmann gas. Although it is hard to put one's finger on precisely what this difference may be, words like "organization", "autonomy", and "top-down causation" come to mind. The issue is well illustrated by considering DNA. This molecule is famously the informational database for an organism's genetic instructions, and the information content may be quantified in bits in a straightforward way. However, any random molecular sequence of the same length would contain a roughly equivalent number of bits, so information content per se does not capture the essence of what makes DNA special. Ask what is the role of a gene, and the answer is: a coded set of instructions for a ribosome to make a protein. Instructional information is clearly more than a mere bit string. There has to be a molecular milieu that can interpret and act on those instructions. In other words, biological information is contextual (perhaps semantic is a better word), i.e. relevant or potent only in the context of a special molecular environment. Context is manifestly a global property. You cannot tell by looking at the local level whether this or that base pair in DNA is instructional information or just junk. So the emergence of life has something to do with the emergence of contextual information processing, a very peculiar sort of complexity. Nobody knows how such a system came to exist even at the conceptual level, nor what it takes in the way of physical processes to bring it about. We certainly understand many mechanisms of physical complexification, but so far none that delivers contextual information processing from the random churning of non-contextual bits.

Because we are ignorant of how life began, it is impossible to estimate the probability of such an event occurring. It may be that life on Earth is a bizarre fluke, a chemical accident of such staggering improbability that it is unlikely to have happened twice in the observable universe. That was the prevailing view fifty years ago. For example, in his book *Chance and Necessity*, Jacques Monod wrote, "The universe is not pregnant with life . . . Man at last knows that he is alone in the unfeeling immensity of the universe, out of which he emerged only by chance" (Monod, 1972). Francis Crick agreed: "Life seems almost a miracle, so many are the conditions necessary for it to get going" (Crick, 1982). In other words, according to Monod and Crick, life is just an exceedingly rare statistical fluctuation. Today the mood has shifted markedly. Thus Christian de Duve writes that life is almost bound to occur wherever Earth-like conditions prevail. "Life is a cosmic imperative!" he declares (de Duve, 1995). Many similar statements are frequently made by distinguished scientists, to the effect that the universe is teeming with life. At this time we have no evidence whatsoever of any life beyond Earth, so that on empirical grounds we cannot discriminate between Monod's view and de Duve's view, between life as a bizarre ultra-rare freak accident and life as a cosmic phenomenon built into the nature of the universe. However, if the latter view is correct, then the story of life's emergence defines a type of complexity arrow of time, because the upward progress from non-life (random informational bits) to life (contextual information processing) is postulated to be a fundamental and universal phenomenon – a "cosmic imperative". Of course, the source of this abiogenesis arrow of time remains a deep mystery, but the emergence of self-reference (Goldenfeld and Woese, 2011) and top-down causation (Davies, 2012) seems to be closely connected with it.

The popular program known as SETI – the search for extraterrestrial intelligence – is predicated on the assumption that there is somehow an inbuilt universal arrow of time from non-life to life (and indeed from life to intelligence), and the success of SETI would confirm that such an arrow of complexity does indeed exist. But detecting a message from aliens may take a very long time, even if such aliens exist, so it makes sense to ask whether the "cosmic imperative" can be tested in some other way meanwhile. An obvious way to test whether the transition from non-life to life is probable is to find a second sample of life, that is, a form of life that has emerged from scratch independently of known life. The best hope for such a discovery is to see whether life has started many times on Earth, by seeking evidence for a "shadow biosphere" in the form of microbes of radically different biochemical structure intermingled with the bacteria and archaea that belong to our own tree of life (Davies and Lineweaver, 2005; Davies *et al.*, 2009; Davies, 2010).

2.6 CANCER AS A CASE STUDY IN ADVANCING COMPLEXITY

Many of the foregoing difficulties are well illustrated by the phenomenon of cancer. Cancer is a remorselessly progressive condition that arises in complex organisms when somatic cells evade the controlling signals of the organism and embark on their own developmental agenda within the host. The clinical hallmarks of cancer are well described (Hanahan and Weinberg, 2000). Cancer cells proliferate uncontrollably to form tumors, evade apoptosis (programmed cell death), flourish in hypoxic conditions, organize their own blood supply, disable tumor suppressor mechanisms including the immune system, secrete special proteins that dissolve the basement membranes near the primary tumor, escape to invade the lymphatic and vascular systems, circulate around the bloodstream, extravasate into tissues very different from their original cell type, co-opt and alter healthy cells there, chemically change the extra-cellular matrix, and grow secondary tumors (metastasis) in the colonized organ. It is important to note that cancer is not a degenerative disease: on the contrary, the neoplasm (the population of new cells) flourishes, at the expense of the host of course. Thus the advance of malignancy defines in a striking (and deadly) manner an arrow of time that correlates with a rise in complexity.

First consider phenotypic complexity. Cancer is often described as a collection of out-of-control rogue cells, a disorderly proliferation. In other words, a neoplasm often appears to be less complex, or at least less organized, than a population of the original somatic cell type. On the one hand, although a tumor lacks the exquisite internal organization of an embryo, it is not an arbitrary clump of cells; rather, it has some features of an ecosystem. Furthermore, there is evidence that the spread of a neoplasm around the body involves a degree of cooperation between cancer cells, or between the migrant cells and the primary tumor. For example, a tumor may "prepare the ground" of a target organ to improve the prospects of a metastatic cell making a home there. This is the so-called seed-and-soil hypothesis (Paget, 1889). And angiogenesis (the creation of a tumor blood supply) also involves a degree of cooperation. In addition, neoplasms are not populations of clones: there is considerable heterogeneity, a factor that contributes to cancer's notorious resistance to chemotherapy. Thus neoplasms are more complex, but less organized, than a population of somatic cells. The disorganization can result in tumor masses of great and growing morphological complexity, resembling fractals. The individual cells also become more complex morphologically, in some cancers displaying a large variety of shapes and sizes. Cancer cell nuclei often adopt grotesque shapes (Nandakumar *et al.*, 2010), while the chromatin they contain is rearranged in complicated ways. Thus the cells themselves become more complex as a function of cancer progression.

The really striking changes associated with cancer are observed, however, at the genotype level. Cancer cells are notoriously genetically unstable, proliferating into a bewildering variety of genotypes.

In well-progressed cancer, the genomes resemble a junk-yard of DNA sequences, with a vast number of alterations. The changes can be at the base-pair (point mutation) level, or the wholesale rearrangement of the chromosomes, with transpositions, relocations, excisions, and duplications of large chunks. Some cancer cells display aneuploidy, in which entire chromosomes may be replicated, perhaps more than once. And yet these "monster cells" continue to function and, to some extent, to cooperate. There is thus an explosive unidirectional growth of genomic complexity associated with cancer progression.

So is cancer in some sense a more advanced, more complex, more varied, form of life, albeit one that has severe consequences for the host organism? Is the originating tissue of the host a stepping stone for a new evolutionary leap to greater variety and complexity? Not according to the fashionable view that cancer cells are somehow a reversion, perhaps to the age of single-celled eukaryotes that prevailed on Earth over one billion years ago. In that sense, they are throwbacks to a simpler evolutionary past pre-dating complex multicellular life. Collectively, a neoplasm may be more diverse and complex than the somatic cells of the originating organ, but individually the cancer cells are not significantly more complex than somatic cells, at least as measured in terms of genetic information content.

The "throwback" theory requires nuancing, however (Davies and Lineweaver, 2011). Contemporary single-celled eukaryotes such as yeast or amoebae do not display the genomic chaos of cancer cells, and it is unlikely that ancient single-celled eukaryotes did either. The well-known ability of cancer cells to flourish, and resist not only multiple layers of defense that the body mobilizes – ranging from tumor suppressor genes, through apoptosis, to the immune system – but in addition much of what the medical profession administers, suggests a highly organized, highly evolved response to proliferative opportunity. This is where the definition of complexity is crucial. According to Davies and Lineweaver (2011), cancer is driven by a highly conserved, highly evolved, cassette of ancient genes stemming from the dawn of multicellularity, and which, although active in early-stage embryogenesis, are normally silenced in the adult form. These genes may be re-awakened as a result of some form of stress or insult, such as radiation or exposure to a carcinogen, and then set about constructing within the body of the host forms that recapitulate ancient (and somewhat primitive) varieties of multicellular life. The hallmarks of cancer are then deployed in a systematic and organized manner, even as the cancer cells abandon control of the remaining genome as superfluous to their agenda, with the concomitant genomic chaos in that inessential part of the chromosomes. There is thus a continuing advance in overall genomic complexity, but bifurcating into an increasingly organized component that drives the cancer and an increasingly chaotic component that results from loss of control over the "irrelevant" (as far as cancer progression is concerned) parts of the genome.

2.7 CONCLUSION

I have argued that there is a widespread belief that many forms of complexity follow a natural upward trend, but that in most cases the scientific foundation for such a trend is shaky. On a cosmological scale it seems that complexity does follow a rising trajectory, but the ultimate end state of the universe may be simple. Similarly, biological complexity, both genotypic and phenotypic, may trend upward, but can go the other way, sometimes dramatically (as after mass extinctions). Evidence for a systematic directionality in biology is not well founded, but may exist, for example, in the encephalization quotient of hominids and in cancer progression. In all cases, precise mathematical definitions of complexity must precede any attempt at arriving at a definitive conclusion.

It may be that there are directionality principles at work in the universe, but that complexity is the wrong concept to focus upon. The real question is whether the universe optimizes something as it evolves, and if so, what. Freeman Dyson has suggested (Dyson, 1979) that the laws of physics are such as to progressively make the universe more and more "interesting", which reflects an intuitive feeling that, over time, the richness, variety and potential of physical systems grows. Identifying just what is meant by "richness" is hard.

It is possible that some mathematical property of the under-lying *laws* of physics is optimized, and that the states of the world evolve in a way that reflects that law-like optimality rather than any optimality in the intrinsic property of the states themselves. Leib-niz argued that we live in the best of all possible worlds, not in any ethical or human sense, but in the sense the universe evolves the greatest complexity in its physical states from the simplest possible underlying laws – a sort of "maximum complexity bang for the buck" principle (Leibniz, 1697). All these musings address the core issue and mystery of existence, which is that the universe is not just "any old" physical system, but one that is in some manner fine-tuned for life, mind, and comprehension (Davies, 2008). Whether a rigorous state-ment of this vague but plausible property can be captured through a definition of complexity and its progressive advancement was the key question that lay before us at the symposium (see the end of Chapter 1).

Acknowledgements

The work reported here was supported in part by NIH grant number U54 CA143682.

REFERENCES

Adams, F. C. & Laughlin, G. (2000). *The Five Ages of the Universe: Inside the Physics of Eternity.* New York: The Free Press.

Atkins, P. (1986). Time and dispersal: the second law. In Flood, R. & Lockwood, M. (eds.), *The Nature of Time.* Oxford: Blackwell.

Atkins, P. (2010). *The Four Laws of Thermodynamics: a Very Short Introduction.* Oxford: Oxford University Press.

Bekenstein, J. D. (1973). Black holes and entropy. *Physical Review, D,* **8**: 2333–2350.

Bennett, C. H. (1988). Logical depth and physical complexity. In Herken, R. *The Universal Turing Machine – a Half-Century Survey.* USA: Oxford University Press, pp. 227–257.

Bousso, R. (2002). The holographic principle. *Reviews of Modern Physics,* **74** (3), 825–874.

Bruner, E., Manzi, G., & Arsuaga, J. L. (2003). Encephalization and allometric trajectories in the genus Homo: evidence from the Neandertal and modern lineages. *PNAS*, **100**, 15335–15340.

Caldwell, R. R., Kamionkowski, M., & Weinberg, N. N. (2003). Phantom energy and cosmic doomsday. *Phys. Rev. Lett.*, **91**, 071301–1.

Chaitin, G. (1987). *Information, Randomness and Incompleteness*. Singapore: World Scientific.

Crick, F. (1982). *Life Itself: Its Nature and Origin*. New York: Simon & Schuster.

Darwin, C. (1863). Letter from Charles Darwin to J. D. Hooker. In Burkhardt, F., Porter, D., Deon, S. A., Topham, J. R., & Wilmot, S. (eds.), *The Correspondence of Charles Darwin 1863*. Reprint (2000) (11): 278. Cambridge: Cambridge University Press.

Davies, P. C. W. (1975). *The Physics of Time Asymmetry*. Berkeley: University of California Press.

Davies, P. C. W. (1977). The thermodynamic theory of black holes. *Proc. Roy. Soc.*, **A 353**, 499.

Davies, P. C. W. (1988). Cosmological event horizons, entropy and quantum particles. *Annales de l'Institut Henri Poincaré*, **49**, 3, 297.

Davies, P. C. W. (1994). *The Last Three Minutes*. London: Weidenfeld & Nicolson.

Davies, P. C. W. (2003). *The Origin of Life*. London: Penguin.

Davies, P. C. W. (2008). *The Goldilocks Enigma: Why is the Universe Just Right for Life?* London: Penguin.

Davies, P. C. W. (2010). *The Eerie Silence: Are We Alone in the Universe?* London: Penguin.

Davies, P. C. W. (2012). Epigenetics and top-down causation. *Interface Focus*, **2**, 42–48.

Davies, P. C. W. & Davis, T. M. (2003). How far can the generalized second law be generalized? *Foundations of Physics*, **32**, 1877–1892.

Davies, P. C. W., Davis, T. M., & Lineweaver, C. H. (2003). Black hole versus cosmological horizon entropy. *Classical and Quantum Gravity*, **20**, 2753–2775.

Davies, P. C. W. & Lineweaver, C. H. (2005). Searching for a second sample of life on Earth. *Astrobiology*, **5**, 154–172.

Davies, P. C. W., Benner, S. A., Cleland, C. E., Lineweaver, C. H., McKay, C. P., & Wolfe-Simon, F. (2009). Signatures of a shadow biosphere. *Astrobiology*, **9**, 241–249.

Davies, P. C. W. & Lineweaver, C. H. (2011). Cancer tumors as Metazoa 1.0: tapping genes of ancient ancestors. *Physical Biology*, **8**, 1–7.

De Duve, C. (1995). *Vital Dust*. New York: Basic Books.

Dyson, F. J. (1979). *Disturbing the Universe*. New York: Harper & Row.

Egan, C. A. & Lineweaver, C. H. (2010). A larger estimate of the entropy of the Universe. *Astrophysical Journal*, **710**, 1825–1834.

Goldenfeld, N. & Woese, C. (2011). Life is physics: evolution as a collective phenomenon far from equilibrium. *Ann. Rev. Condens. Matter Phys.*, **2**, 17.1–17.25.

Gould, S. J. (1996). *Full House: the Spread of Excellence from Plato to Darwin*. New York: Harmony Books.

Hanahan, D. & Weinberg, R. (2000). The hallmarks of cancer. *Cell*, **100**, 57–70.

Hawking, S. W. (1971). Supermassive objects in astropysics. *Phys. Rev. Lett.*, **26**, 1344.

Hawking, S. W. (1975). Particle creation by black holes. *Communications in Mathematical Physics*, **43**, 199–220.

Hawking, S. W. (1978). Space-time foam. *Nucl. Phys.*, B, **144**, 349.

Leibniz, G. W. (1697). *Philosophical Writings*. Translated (1997) by Morris, M. and Parkinson, G. H. R. Edited by Tuttle, C. E. Vermont: Everyman.

Lineweaver, C. H. & Egan, C. A. (2008). Life, gravity and the second law of thermodynamics. *Physics of Life Reviews*, **5**, 225–242.

Lloyd, S. & Pagels, H. (1988). Complexity as thermodynamic depth. *Ann. Phys.*, **188**, 186.

Monod, J. (1972) *Chance and Necessity*. Translated by Wainhouse, A. London: Collins, p. 167.

Nandakumar, V., Kelbauskas, L., Johnson, R., & Meldrum, D. (2010). Quantitative characterization of preneoplastic progression using single-cell computed tomography and three-dimensional karyometry. *International Society for Advancement of Cytometry*, **79A**, 25–34.

Nitecki, M. H. (1989). *Evolutionary Progress*. Chicago: University of Chicago Press.

Paget, S. (1889). The distribution of secondary growths in cancer of the breast. *Lancet*, **133**, 571–573.

Penrose, R. (1979). Singularities and time-asymmetry. In Hawking, S. W. & Israel, W. *General Relativity: an Einstein Centenary Survey*. New York: Cambridge University Press, pp. 581–638.

Russell, B. (1957). *Why I Am Not a Christian*. New York: Allen & Unwin.

Thompson, W. (1852). On a universal tendency in nature to the dissipation of mechanical energy. *Proceedings of the Royal Society of Edinburgh*, April 19, 1852.

3 A simple treatment of complexity: cosmological entropic boundary conditions on increasing complexity

Charles H. Lineweaver

3.1 THE COMPLEXITY OF COMPLEXITY

Proud Biologist: "Life forms are more complex than stars"
Humble Astronomer: "You'd look simple too from a trillion miles away"

One of the central questions of evolutionary biology and cosmology is: is there a general trend towards increasing complexity? In order to answer that question, it would help to have a definition of complexity that can be quantified. Various definitions of complexity have been proposed (Gell-Mann, 1994, 1995; Kauffman, 1995; Adami, 2002; Gell-Mann & Lloyd, 2003; Fullsack, 2011). With useful oversight, Lloyd (2001) groups various conceptions of complexity into three groups based on (1) difficulty of description (measured in bits) (2) difficulty of creation (measured in time, energy or price) and (3) degree of organization (measured in . . . ? . . . , we're not sure). For more details see: Weaver, 1948; Traub *et al.*, 1983; Chaitin, 1987; Weber *et al.*, 1988; Wicken, 1988; Bennett, 1988; Lloyd & Pagels, 1988; Zurek, 1989; Crutchfield & Young, 1989; McShea, 2000; Adami *et al.*, 2000; Adami, 2002; Hazen *et al.*, 2008; Li & Vitanyi, 2008; McShea & Brandon, 2010.

I will not try to unify these mildly-compatible definitions of complexity, since such an effort would probably resemble the

Complexity and the Arrow of Time, ed. Charles H. Lineweaver, Paul C. W. Davies and Michael Ruse. Published by Cambridge University Press.
© Cambridge University Press 2013.

confusing attempts to one-dimensionalize the N-dimensional concept of human intelligence. However, a unifying feature of the effective complexity discussed here (Gell-Mann & Lloyd, 2003) is the intuitive notion that complexity lives somewhere in the continuum between complete order and random chaos (Crutchfield & Young, 1989; Gell-Mann, 1995; Adami, 2002). Complex systems are far-from-equilibrium-dissipative systems (Prigogine, 1978). Thus, we are far from equilibrium (i.e. far from random) but we are also far from order. If simplicity is the opposite of complexity, then both random chaos and complete order are simple. In a cosmological context, order can be understood as the low entropy condition of homogeneously distributed matter in the early universe.

The fact that life forms are complex and that our DNA contains information about the environment (past and present) seems obvious and has been emphasized by various authorities:

[Organisms] encode the predictable occurrence of nature's storms in the letters of their genes.

(Wilson, 1992)

genes embody knowledge about their niches.

(Deutsch, 1997)

Adami *et al.* (2000) identify genomic complexity with the amount of information DNA sequences store about their environments. I want to emphasize that not only is the information in DNA *about* the environment, but that the information in DNA *came* from the environment (see Krakauer, Chapter 10). In any naturalistic explanation for the origin and evolution of life, the non-adaptive complexity of the physical environment precedes, and is the source of, the adaptive complexity of life. The information in DNA comes from the environment and it has been put into the DNA by selection. Darwinian evolution involving selection of all kinds (natural, sexual, and artificial) is the channel through which the complexity and information of the environment creates and shapes the complexity and information of biological organisms (e.g. Spiegelman, 1971).

The complexity of the environment is in the spatial and temporal differences of such variables as temperature, density, pressure, chemistry, and the availability of energy, water, and nutrients. These detailed non-adaptive structural complexities and differences did not always exist (Zaikowski & Friedrich, 2008). These differences started out as small fluctuations about equilibrium and were amplified by gravitational collapse. Galactic clouds of hydrogen evolved into stars. Stars evolved layers and produced complicated patterns of isotopic abundances in the interstellar medium. Undifferentiated objects in proto-planetary disks irreversibly differentiated and became planets, with density-segregated layers (core/mantle/crust/oceans/atmospheres) whose surfaces are pockmarked information-rich palimpsests of the history of the Solar System. The number of minerals on Earth has increased with time (Hazen *et al.*, 2008; Hazen & Eldredge, 2010).

A subset of these differences provides useful gradients from which free energy can be extracted – gradients of luminosity, redox chemistry, pH, temperature, humidity, density, gravity, etc. (Schneider & Sagan, 2006; Lineweaver & Egan, 2008, 2011). The maintenance of irreversible processes requires the dissipation of these gradients and their associated free energy (Ulanowicz & Hannon, 1987). Or, equivalently, the flow of free energy driven by these gradients produces dissipative structures which are maintained as long as the gradients persist (e.g. Lineweaver & Egan, 2008; Kleidon, 2012).

Various forms of irreversible processes and dissipative structures produce and maintain complexity. All forms of irreversible processes are subject to entropic boundary conditions – these include simple near-equilibrium structures such as cooling planets and cups of coffee, but there are also far-from-equilibrium dissipative structures of varying complexity such as convection cells, hurricanes, and life forms (Prigogine, 1978). We are most interested in life forms since that is what we are. We have a mechanism (inheritable coded molecules of DNA or RNA) for storing information about the environment inside ourselves and passing it on to descendants. Thus we are *adaptive* systems. Hurricanes don't do that. Non-biological far-from-equilibrium dissipative structures don't do that. They are *non-adaptive* systems. However, the limits on the availability of free energy are limits for all irreversible structures – both non-adaptive and adaptive – from hot coffee cups and hurricanes to life forms.

In summary, we are interested in the evolution of complexity in the universe. Just as biological complexity can be traced back to the complexity of the environment, environmental complexity can be traced back to free energy available due to an entropy gap between the initial low entropy of the universe and the maximum potential entropy of the universe. Complexity is limited by the availability of free energy and free energy is limited by the entropy gap.

3.2 EVOLUTION OF THE ENTROPY AND THE MAXIMUM POTENTIAL ENTROPY OF THE UNIVERSE

There is general agreement that the entropy of the universe ($"S_{uni}"$ in Fig. 3.1), started out low and has been increasing ever since (Penrose, 1979, 2004; Davies, 1994; Lineweaver & Egan, 2008, 2011; Egan & Lineweaver, 2010; Carroll, 2010). All three panels in Fig. 3.1 have S_{uni} starting out low. An initial low entropy is not obvious since observations of the cosmic microwave background (Smoot *et al.*, 1992) revealed the conditions of the universe $\sim 400\,000$ years after the big bang. They revealed a nearly homogeneous, isotropic, isothermal, isobaric, iso-everything universe (at least at the level of a few parts in 10^5). There were no stars or planets or galaxies. These observations seem to suggest not a low initial entropy but a high initial entropy – a universe in equilibrium at its maximum possible value, S_{max}. This apparent equilibrium of the early universe is why $S_{uni} = S_{max}$ at early times in Fig. 3.1(a) & (b). But if the universe started out in equilibrium with $\Delta S = 0$, how did anything happen? What drove it out of equilibrium? Fig 3.1(c) does not have this problem because it includes in S_{uni} the low initial gravitational entropy of nearly homogeneously

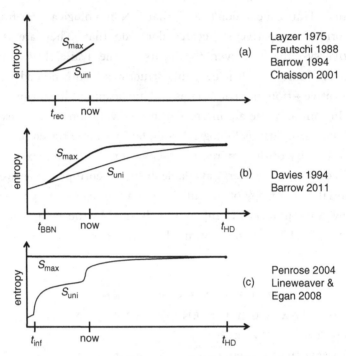

FIGURE 3.1 Three different views of the evolution of the entropy of the universe, S_{uni}, and the maximum potential entropy of the universe, S_{max}. In (a), t_{rec} is the time of recombination. In (b), t_{BBN} is the time of big bang nucleosynthesis. In (c), t_{inf} is the time of inflation. In (b) & (c), t_{HD} is the heat death of the universe. We would like to understand why these sketches are so different and which (if any) gives the most qualitatively correct picture. One important difference is what to include in S_{max} and S_{uni} (see text).

distributed matter. This gravitational entropy has been ignored in Fig. 3.1(a) & (b). Also, in Fig. 3.1(c), S_{max} is defined differently.

The entropy gap,

$$\Delta S = S_{max} - S_{uni} \tag{3.1}$$

shown in Fig. 3.2 is the difference between the maximum potential entropy and the actual entropy of the universe (Lineweaver & Egan, 2008). In Fig. 3.1, all three panels agree that both S_{uni} and the maximum potential entropy S_{max} cannot decrease – both obey

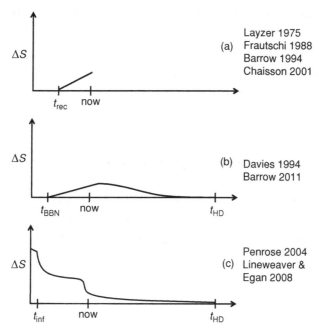

FIGURE 3.2 Same as Fig. 3.1 except the y-axis is now the difference $\Delta S = S_{max} - S_{uni}$ (Eq. (3.1)) from each panel of Fig. 3.1. ΔS is important because it is a measure of free energy (Eq. (3.4)), which is the only thing that can maintain existing complexity or drive increasing complexity. Panel (a) suggests that ΔS continues to increase indefinitely, thus potentially allowing for an unlimited increase of free energy and complexity. In panels (b) and (c), $\Delta S \to 0$ at the heat death of the universe, suggesting the dissipation of all free energy and the reduction and disappearance of complexity. The significant disagreement between panel (a) and the other two needs to be resolved if we are to resolve the long term fate of the complexity of the universe.

the second law of thermodynamics. In panel (c), S_{max} is a constant set at our estimated value of the highest entropy the universe will ever reach, through known dissipative processes such as black hole formation and evaporation (Egan & Lineweaver, 2010). However, in panels (a) and (b), S_{max} represents a time-dependent maximum potential entropy that *would* be produced if all the free energy at time t could somehow be dissipated through instantaneous and unknown dissipation mechanisms into the coldest heat sink available at time t.

The reason S_{max} continues to rise in panel (a) is discussed in the next section.

The Helmholtz free energy is $F = U - TS$ (e.g. Bejan, 2006). With constant energy U and steady state temperatures, we can write any change in the available free energy as

$$dF = -TdS,$$ (3.2)

i.e. when entropy increases, the amount of free energy decreases (Lineweaver & Egan 2008, 2011). Since this is a simple treatment of complexity, the controversial caveats about applying the equations of thermodynamics to non-equilibrium conditions are ignored (e.g. Jaynes, 1989; Kleidon & Lorenz, 2005; Rubi, 2008). Thus,

$$\int_{F_{max}}^{F_{min}} dF = -T \int_{S_{uni}}^{S_{max}} dS,$$ (3.3)

$$\Delta F = T \cdot \Delta S.$$ (3.4)

The entropy gap ΔS is a measure of the amount of free energy left in the universe. Free energy is the only kind of energy able to drive irreversible processes such as life and any increase in complexity. Whatever complexity is, it cannot increase without a supply of free energy since all forms of complexity involve irreversible processes. Just as life (or biological complexity) depends on the free energy available from an environment out of equilibrium (Schrödinger, 1944), the increase of any kind of complexity cannot happen without a supply of free energy. The availability of free energy is a necessary but not sufficient requirement for complexity since ΔF and ΔS are not measures of complexity, but are measures of the potential for complexity.

> Thermodynamic potentials such as entropy or free energy measure capacity for irreversible change, but do not agree with subjective complexity. A human body is more complex than a vat of nitroglycerine, but has lower free energy.
>
> (Bennett, 1994)

The complexity of animal bodies has more to do with the evolved complexity associated with its ability to tap into a flow of free energy than with the free energy value of their contents. It is hard to say more about the relationship between free energy flow and complexity without a definition of complexity.

As $\Delta S \to 0$ (as it does in Fig. 3.2(b) & (c)), we have $\Delta F \to 0$, and therefore all food for life and the ability to maintain complexity goes to zero. We are not suggesting that large ΔS and large ΔF are equivalent to high complexity. The low-complexity early universe with large ΔS (according to Fig. 3.1, panel (c)) is probably the best example of how a large entropy gap is not a good measure of complexity.

Whatever measure of complexity we use, there was little of it in the first tens of millions of years after the big bang when ΔF was large (Fig. 3.1(c)). The first stars formed only after a few hundred million years and the first terrestrial planets formed about a billion years later (Lineweaver, 2001). Thus the early universe was not complex according to any current definition of complexity.

A non-zero entropy gap ΔS is a necessary, but possibly not a sufficient, condition to produce complexity. If it is necessary *and* sufficient, then there must be a considerable time lag before available free energy produces complexity. This is plausible since it takes time for power to spread through the system from primary sources of free energy to secondary sources. This concept is central to Bennett's (1988) slow growth law, under which it takes time for a low entropy universe to evolve the dissipative systems that produce complexity. Complex adaptive systems have a tendency to give rise to other complex adaptive systems (Gell-Mann, 1995), but this requires dissipation, the decrease of free energy, and time. The evolution of the complexity of non-adaptive systems also takes time. For example, it takes a few hundred million years for star formation to access the free energy of nuclear potential of hydrogen. And it takes millions of years of accretion for black holes to access the dominant amount of the gravitational potential of the more homogeneously distributed matter around it. For examples of how solar power spreads through the subsystems of the Earth, see Kleidon (2010) and Lineweaver (2010).

3.3 EXISTING EQUILIBRIUM, OPENING AN ENTROPY GAP, AND THE CONCEPTUAL PROBLEM OF THE MAXIMUM POTENTIAL ENTROPY OF THE UNIVERSE

The central conceptual problem in Fig. 3.1(a) & (b) is this: once the universe is at equilibrium, what processes can get it out? Equilibrium is a state around which the universe can fluctuate (Evans & Searles, 1994), but the universe cannot ratchet its way out of equilibrium without violating the second law. Based on the second law I argue that if it appears that an entropy gap is opening, then that is because there is some unrecognized free energy that should have been included in S_{max} but was not. In Fig. 3.1(a) & (b), Layzer (1975), Frautschi (1988), Barrow (1994, 2011), Chaisson (2001), and Davies (1974, 1994), start the universe in equilibrium, at maximum entropy, with $S_{uni} = S_{max}$. The process that is claimed to open the entropy gap in Fig. 3.1(a) & (b) is the expansion of the universe driving the universe out of equilibrium as S_{max} increases faster than S_{uni}. As the universe expands and cools, components that had once been in thermal equilibrium with each other fall out of thermal equilibrium with each other. For example, gravitons decoupled at the Planck time from the rest of the universe. Two seconds later, neutrinos decoupled from the cosmic background radiation, such that today, gravitons, neutrinos, and cosmic background photons co-exist at different temperatures: ~ 0.6 K, 1.95 K and 2.7 K respectively (Egan & Lineweaver, 2010). However, this does not open an entropy gap ΔS that can be inserted into Eq. (3.4) because no work can be extracted from the decoupled fluids.

In Fig. 3.1(a), consider the opening of the entropy gap at recombination. The reason for this opening is supposed to be the emergence of a temperature difference between photons and matter. Such a temperature difference can arise in two ways. One way is that, after recombination, the temperature of the photons scales as $1/a$ while the temperature of the matter scales as $1/a^2$ (where a is the scale factor of the universe). Thus, as a increases, the temperature of the matter cools faster than the temperature of the radiation – the temperatures of the photons and matter diverge. Temperature differences in general are often associated with the ability to drive winds or convection cells and do work (e.g. steam engines or internal combustion engines). However, this cosmological photon/matter temperature difference is between two *decoupled, non-interacting fluids*, both of which are ubiquitous and co-spatial. No work or free energy can be extracted from the temperature difference of decoupled fluids any more than Maxwell's demon can extract work by spatially separating fast and slow particles. No winds or convection cells are produced. Analogous statements can be made about any two decoupled, co-spatial fluids such as between the present 2.7 K cosmic microwave background and the 1.95 K neutrino background – or between either of these and the < 1 K graviton background. Thus, the expansion of the universe and the decoupling of matter from photons cannot open an entropy gap that could be a source of free energy in Eq. (3.4). However, the decoupling of matter and photons is not instantaneous nor complete. A small amount of heat can flow from the hotter photons to the cooler matter because of this residual coupling. This produces some entropy and is known as bulk viscosity (e.g. Zimdahl & Pavon, 2001). But because the hot and cold are not separated by any macroscopic boundary, that is, because there is no macroscopic temperature gradient, no work can be extracted and no dissipative structure can form.

The other way to open up a temperature difference between matter and photons is to heat the matter with an extra source of UV photons, ionizing the matter to temperatures $\sim 10^4$ K. This is what happened during the epoch of re-ionization at a redshift $z \sim 12$. However, the free energy source of the photons was the gravitational accretion energy from either active galactic nuclei or shocks (Dopita *et al.*, 2011) or nuclear fusion in population III stars. This free energy existed before the temperature difference. In other words, some previously existing source of free energy ($\Delta F > 0$ and $\Delta S > 0$) drove re-ionization and the temperature difference. It was not the expansion of the universe or the increase of the scale factor a.

Beyond the entropy produced from bulk viscosity, the expansion of the universe does not increase the entropy of a comoving volume of relativistic energy or non-relativistic matter (Lineweaver & Egan, 2008; Egan & Lineweaver, 2010). Nor does the expansion of the universe increase the entropy of the universe when one species of particle of mass m decouples at equilibrium ($mc^2 \sim kT$), even though at a later epoch, as the photon temperature T decreases, we have $mc^2 > kT$. It is only if the massive particle is unstable and decays under the conditions $mc^2 > kT$ that we have an entropy increase attributable to the expansion of the universe. But if these unstable particles are homogenous and microscopically mixed with all the other particles, no work can be extracted.

Consider the opening and closing of the entropy gap in Fig. 3.1(b). Davies (1994) wrote about it:

> It is important to realize that the crucial effect of the expansion was in the early universe – hence the sudden widening of the gap early on. Today it seems likely (though I haven't checked) that the gap is narrowing: the universe produces copious quantities of entropy at a rate which I imagine is faster than the (now rather feeble) expansion raises the maximum possible entropy. The actual entropy will presumably asymptote towards the maximum possible entropy in the very far future.

The idea behind the opening of the gap after t_{BBN} in Fig. 3.1(b) is that before t_{BBN} the temperature is too high for free-energy-yielding nuclear fusion to occur. For example, nuclear fusion cannot produce free energy when the entire universe is a quark–gluon plasma. The first three minutes was not hot enough or dense enough or long enough (due to the rapid expansion of the universe) to complete fusion and release all potential nuclear binding energy. Big bang nucleosynthesis (BBN) and the expansion of the universe left lots of hydrogen, deuterium, and helium that had not been burned to iron (cf. Fig. 4 of Lineweaver & Egan, 2008). If heated up later in the cores of stars at high density, hydrogen can fuse into heavier elements and eventually into iron. The view taken in Fig. 3.1(b) is that before t_{BBN}, this potential nuclear free energy should not be included in the entropy gap, but that after t_{BBN} this potential nuclear free energy should be included. One could argue that in order for nucleosynthesis to open up an entropy gap that could be associated with free energy and work, the hydrogen has to have collapsed into stars and this doesn't happen until a few hundred million years after the big bang. The time at which the "potential" for such gravitational collapse appeared is not well defined. It could be before t_{BBN} when an excess of matter over antimatter appeared in the universe. Or it could be when the universe transitioned from radiation dominated to matter dominated, allowing cold dark matter to collapse to form the seeds of large scale structure. Or it could be at recombination, when baryonic matter decoupled from photons and began to clump in the over-dense cold dark matter haloes. Or it could be a few hundred million years later when hydrogen began to fuse into helium in the first stars. Contrafactual "potential" is a slippery concept.

Similarly, arbitrary S_{max} budgeting produces similar confusion with the accounting of the entropy of black holes. As black holes form, the entropy of the universe S_{uni} increases. But when and how are we to include potential black hole formation into the budget of S_{max}? What does it mean to include in S_{max} the entropy of black holes that *could* form, but *never will form*? There are well-discussed entropy bounds in the literature, for example, which envisage all the matter in the observable universe collapsing into a black hole (Susskind, 1995). But we live in a Λ-dominated universe in which the acceleration of the universe has been shutting off the growth of structure for the past billion years. Thus, the entropic bound of all matter in the observable universe collapsing into a black hole cannot give us an attainable or plausible value for S_{max}. If we are concerned with values of S_{max} that can be inserted into Eqs. (3.1) and (3.4) to give us the currently most plausible estimate of the amount of free energy that eventually becomes available to produce complexity during the evolution of the universe, then the most meaningful S_{max} seems to be the one in Fig. 3.1(c).

Another reason why (contrary to what is shown in Fig. 3.1(a) & (b)) S_{uni} cannot be equal to S_{max} for all $t < t_{BBN}$ is that the asymmetry between matter and antimatter ("baryogenesis") had its origin in conditions that required thermodynamic disequilibrium or $\Delta S > 0$ (Sakharov, 1967; Kolb & Turner, 1990; Quinn & Nir, 2008).

With the establishment in $\sim$ 1998 of the cosmological constant as the dominant form of energy in the universe, Barrow's (1994) views changed from Fig. 3.1(a) to Fig 3.1(b) (Barrow, 2011, personal communication). Also, in contrast with Davies (1994), Barrow sees the entropy gap opening at the Planck time, about 3 minutes ($\sim$ 45 orders of magnitude) earlier than at the t_{BBN} shown in Fig. 3.1(b).

3.4 SOURCES OF FREE ENERGY

In the standard ΛCDM big bang model with an early epoch of inflation tacked on at the beginning (e.g. Liddle & Lyth, 2000), the sources of free energy in the early universe are:

(1) vacuum energy. At the end of inflation there was a transition from a false vacuum to a true vacuum. The potential energy of the false vacuum was dumped into the universe in the form of radiation, matter, and antimatter;

(2) the disequilibrium that produced more matter than antimatter (Sakharov, 1967) and is a source of free energy in that, if there were no asymmetry, the universe would contain only radiation and there would be no matter that could collapse;

(3) the gravitational potential energy of the homogeneous distribution of the excess matter.

After the energy of the vacuum was dumped into the universe, this energy was out of equilibrium in at least two ways. Firstly, it had to be in disequilibrium to even produce a matter–antimatter asymmetry. Secondly, homogeneous matter can clump into inhomogeneities. The mutual annihilation of matter and antimatter was a source of energy but was not itself a source of free energy for the same reason that the heat transfer from photons to baryons after recombination was not a source of free energy: no macroscopic boundary between source and sink.

The transition from unclumped matter to clumped matter is still going on today, creating gradients of density, pressure, and temperature, and which, at the center of stars, is permitting access to the unburned nuclear binding energy of hydrogen and helium. All of this makes the entropy of the universe increase. As matter clumps, the entropy increases, but there is not yet an equation linking the parameters of large scale structure formation to gravitational entropy. The high gravitational potential energy and the associated low gravitational entropy of initially unclumped matter is the main fact upon which Penrose (1979, 2004) and Lineweaver & Egan (2008, 2011) claim that the universe started out at low entropy and that initial ΔS was at a maximum. We hypothesize that the low entropy origin of the universe was due to the highly homogeneous matter far from gravitational equilibrium, despite it being near thermal and chemical equilibrium.

One source of the conceptual confusion underlying the differences in the sketches of Fig. 3.1 is whether one should assign a large initial potential S_{max} to this unclumped matter or not. The disagreement is not about how or when matter clumps, but about how much of its not-yet-clumped potential to assign to S_{max} and when to assign it. Should a metastable local minimum of energy be considered equilibrium or disequilibrium (Fig. 3.3)? Trying to estimate ΔS from how much matter could clump now (but hasn't) seems to be a difficult or even unaddressable issue which we circumvent by estimating the ultimate extent to which matter will clump. Thus in Egan & Lineweaver (2010) we chose to use a constant S_{max} set by the degree to which matter eventually clumps (under our current assumptions about the far future of the universe).

Consider Fig. 3.3. If the universe starts out in a false vacuum at ϕ_1 and then tunnels into a true vacuum at ϕ_2, then the potential energy difference $\Delta V_{12} = V_1 - V_2$ is dissipated during reheating and

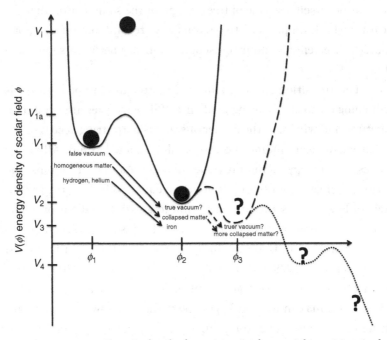

FIGURE 3.3 Generic sketch of transitions and potential transitions in the
early universe that are the sources of free energy ($\Delta V \sim \Delta F$ of Eq. (3.4)). In
order to have a ΔS that increases without bounds (Fig. 3.2(a)) we require an
infinite cascade of phase transitions beyond the supposedly true vacuum
of ϕ_2, similar to the three transitions described between ϕ_1 and ϕ_2, i.e.
false to true vacuum, unclumped to clumped matter, and hydrogen to
iron.

the entropy of the universe goes up. Therefore the universe in the
false vacuum before the end of inflation is in disequilibrium, not at
$S_{uni} = S_{max}$ as is assumed in Fig. 3.1(a) and (b). During reheating, matter was dumped homogeneously into the universe. The matter could
then begin to collapse. Thus, a universe with homogeneously distributed matter is in disequilibrium with respect to gravity, because
the matter has not all collapsed. When matter does collapse and large-
scale structure forms, heat is released and entropy increases. Thus,
our early universe with homogeneously distributed matter was in
disequilibrium, not at $S_{uni} = S_{max}$.

The same reasoning goes for hydrogen and iron. A universe filled with hydrogen and helium is filled with a fuel, at a local minimum of nuclear binding energy (e.g. ϕ_1 of Fig. 3.3). We interpret this local minimum of the energy as a metastable recoverable disequilibrium source of free energy that has existed since very early in the evolution of the universe when the expansion rate and density and therefore the eventual incompleteness of BBN were determined (see Fig. 4 of Lineweaver & Egan, 2008). We consider nuclear potential free energy as being recoverable because it will eventually burn in our universe and be taken out of this local minimum by the high temperatures and densities at the centers of stars. If we lived in a universe in which no stars formed, then hydrogen and helium would not be a source of free energy and we would not include this source of entropy in S_{max}. Thus, the transitions from a false vacuum to a true vacuum, from unclumped matter to clumped matter, and from hydrogen to iron are sources of disequilibrium that exist initially and are the reasons why ΔS in Fig. 3.1 panel (c) is so large.

One disadvantage of the constant S_{max} approach shown in Fig. 3.1(c) is that one would like ΔS to be a measure of how much free energy is available at time t to drive the dissipative processes that are occurring at time t. This is not the case for our constant S_{max} and the ΔS derived from it, which are measures of the free energy that will eventually become available at some time during the evolution of the universe. For example, although nuclear fusion of hydrogen into helium is included in our ΔS at early times (i.e. $t <$ few hundred million years), the first stars, giving access to temperatures and densities which permit fusion, do not exist until a few hundred million years after the big bang.

One way to make the concept of maximum entropy more useful is to distinguish between the constant S_{max} of Fig. 3.1(c) and a time-dependent $S'_{max}(t)$ that depends on the rate at which entropy is being produced. For example, we could define it as

$$S'_{max}(t) = S_{uni}(t) + (\alpha/H)dS_{uni}/dt, \tag{3.5}$$

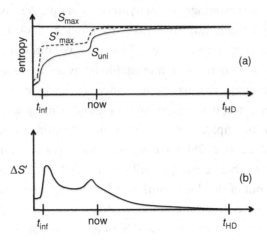

FIGURE 3.4 Same as Figs. 3.1(c) and 3.2(c) except showing $S'_{max}(t)$ in (a) and the resulting new $\Delta S'$ in (b). See Eq. (3.5)–(3.8).

where H is Hubble's constant and the dimensionless constant α is constrained by the condition

$$S_{uni}(t) \leq S'_{max}(t) \leq S_{max}, \qquad (3.6)$$

which ensures that in the far future, as the universe approaches a heat death, that $S'_{max}(t = t_{HD}) = S_{max}$. With this new maximum entropy we can define a new entropy gap,

$$\Delta S' = S'_{max}(t) - S_{uni}(t) = (\alpha/H)dS_{uni}/dt \qquad (3.7)$$

and a modified version of Eq. (3.4),

$$\Delta F' = T\Delta S' \qquad (3.8)$$

that reflects the free energy $\Delta F'$ available as a function of time and depends on the instantaneous rate at which the entropy of the universe is increasing, dS_{uni}/dt. In a cash-flow analogy, Eq. (3.8) amounts to trying to estimate how much a customer has to spend now ($\Delta F'$) by measuring how much they are spending ($\Delta S'$). This is different from the total amount a customer will ever be able to spend (ΔF of Eq. (3.4)). Figure 3.4 sketches what $S'_{max}(t)$ and $\Delta S'$ might look like. Notice that, in contrast to ΔS of Fig. 1(c), $\Delta S'$ increases when free energy becomes available (e.g. as stars fuse hydrogen). In contrast with ΔS of Fig. 3.1(b), $\Delta S'$ does not depend on non-existent instantaneous dissipation mechanisms or subjective estimates of the potential for dissipation. $\Delta S'$ depends on a measurable quantity, dS_{uni}/dt.

3.5 DOES COMPLEXITY INCREASE?

Does complexity increase in the course of evolution? Or does it decrease? According to the Second Law of Thermodynamics the latter should be the case. But looking just superficially at the richness of nature, one comes to believe in an ongoing and open-ended emergence of increasingly complex structures that stabilize further and further from thermodynamic equilibrium – with humans and their creations possibly being the latest manifestations of this.

(Fullsack, 2011)

To answer the question "is complexity increasing?" we need to disambiguate it. Are we talking about the average complexity of the universe or about the complexity of the most complex object? Are we talking about a current increasing trend (that could be quite ephemeral and last for only a million or a billion years)? Or are we talking about an ultimate enduring trend? In this chapter we use the connection between complexity and entropy to conclude that the ultimate trend of the average complexity must be to decrease.

Since all laws of physics are time-reversible except for the second law, if there is a secular change in some quantity, such as complexity, then it will have a deep connection with the only law that has a direction for time, the second law. This is the connection we have used in this chapter.

Complexity relies on $\Delta F > 0$. However, the combination of the cosmological constant and the second law of thermodynamics requires $\Delta S \to 0$ and therefore $\Delta F \to 0$ (Eq. (3.4)). This entails the heat death of the universe, the fading of complexity like a flashlight with a dying battery, the extinction of all life, and the disappearance of all structure, leaving us in the simplicity of equilibrium, forever.

This conclusion seems to be in disagreement with Dyson (1979) but in agreement with Krauss & Starkman (2000). And it leaves room for comforting statements like:

> *[W]e will still ultimately lose the battle against degeneration. But the second law does not mandate a steady degeneration. It quite happily coexists with spontaneous development of order and complexity.*
>
> *(Rubi, 2008)*

The happy coexistence of the second law and complex objects that Rubi is referring to is based on a continuous but unsustainable expulsion of high entropy material by the most complex objects. Even in a universe in which the average complexity is decreasing, the complexity of the most complex objects can increase for a while. This is certainly the niche we identify with, but like burning fossil fuels, it is not sustainable.

Long-term patterns of biological evolution are not exempt from the second law. Adami *et al.* (2000) describe how biological evolution acts like a Maxwell demon. Maxwell's demon (Maxwell, 1888) lowers the entropy of molecules in a box by letting only hot molecules pass from one side to the other – thus separating the hot from the cold molecules, and apparently violating the second law. Natural selection can also be thought of as a Maxwell's demon selecting fitter phenotypes (~ "hotter molecules") whose DNA contains more information about the environment. But just as Szilard (1929) and Bennett (1987) have pointed out that Maxwell's demon is not a perpetual motion machine, the Maxwell demon of natural selection is not a perpetual motion machine. There is a source of free energy that both provides an information-rich environment and the energy that sustains life. Driven by free energy, natural selection turns the crank that ratchets up and preserves the accumulation of information in DNA.

> *Darwinian selection is a filter, allowing only informative mea-*
> *surements (those increasing the ability for an organism to sur-*
> *vive) to be preserved. In other words, information cannot be lost*
> *in such an event because a mutation corrupting the information*
> *is purged due to the corrupted genome's inferior fitness.*
>
> *(Adami* et al.*, 2000)*

We have argued that a complex environment is the result of the initial low gravitational entropy of the early universe and the resulting grav-itational collapse of galaxies, stars, and planets. The ultimate driver of complexity is the dissipation of free energy. This does not seem to be a well-accepted point of view. Gell-Mann (1995) does not associate the evolution of complexity with low gravitational entropy, but wants to explain the complexity of life as the result of "frozen accidents, giving rise to regularities" with little or no connection with free energy.

> *As the universe grows older and frozen accidents pile up, the*
> *opportunities for effective complexity to increase keep accumu-*
> *lating as well. Thus there is a tendency for the envelope of com-*
> *plexity to expand.*
>
> *(Gell-Mann, 1995)*

> *The second law of thermodynamics, which requires average*
> *entropy (or disorder) to increase, does not in any way forbid*
> *local order from arising through various mechanisms of self-*
> *organization, which can turn accidents into frozen ones produc-*
> *ing extensive regularities.*
>
> *(Gell-Mann, 1995)*

Note the similarity between this statement and Miguel Rubi's state-ment above. Both refer to the uncontroversial increase in the local order. However, the crucial ingredient not mentioned in the recipe is that these "extensive regularities" or complexities (like the ΔT produced by Maxwell's demon) come at a price of higher entropy else-where. And since sources of free energy are always decreasing, the trend toward local order and complexity, like a civilization built on fossil fuel, can only be temporary.

Does local order keep increasing? It can until the exported entropy fills up the universe. For example, air conditioners and refrigerators work as long as the heat they generate can be removed . . . as long as there is a sink. Segregation can continue, but will not last forever since the amount of free energy is limited and without free energy there can be no segregation, no export of high entropy, leaving the low entropy behind.

We know that life forms are not unusual statistical fluctuations or Boltzmann brains because we persist in ways that 1000 sigma statistical fluctuations do not. When the molecules in this room pile into a corner at random, they immediately pile out. They are not frozen. The reason regularities can be frozen into life is because of a constant free energy supply which supplies the electricity to the freezer. This persistence requires a flow of free energy whose dissipation is the price of our persistence. You cannot freeze accidents for free.

If we find that we are living in a false vacuum and that protons and other seemingly stable particles decay, then these will be new sources of free energy and the universe will be able to evolve to the right in Fig. 3.3. If an infinite number of such free energy sources are identified then the universe can keep evolving to the right in Fig. 3.3 forever. On the other hand, if we have already identified all the sources of free energy in the universe, then the acceleration of the expansion of the universe and its asymptotic approach to a vacuum state will lead to the heat death of the universe, the dissipation of all free energy and the reduction and disappearance of complexity as shown in Fig. 3.2, panel (c).

3.6 SUMMARY

In any naturalistic explanation for the origin and evolution of life, the non-adaptive complexity of the physical environment precedes, and is the source of, the adaptive complexity of life. However, the complexity of the physical environment did not always exist. 400 000 years after the big bang there were no stars, planets, or life. The complexity of the physical environment is the result of irreversible processes driven by the dissipation of free energy – initially gravitational free energy associated with the initial low entropy of the universe. Since the amount of free energy decreases as the entropy of the universe increases, cosmological estimates of entropy yield upper limits on physical complexity and therefore biological complexity. I used the concept of an entropy gap, $\Delta S = S_{max} - S_{uni.}$, between the maximum entropy and the actual entropy of the universe to quantify the available free energy and the potential for complexity in the universe. Previous estimates of ΔS were compared and found to differ because of different assumptions about S_{max}, equilibrium and free energy. I have clarified some of these differences. I found that the combination of the cosmological constant and the second law of thermodynamics requires $\Delta S \to 0$. This entails the heat death of the universe, the decrease of complexity like the fading glow of a flashlight with a dying battery, the extinction of all life, the disappearance of all structure – leaving us in the simplicity of equilibrium, forever and ever. Amen.

REFERENCES

Adami, C. (2002). What is complexity? *BioEssays*, **24**, 12, 1085–1094.

Adami, C., Ofria, C., & Collier, T. C. (2000). Evolution of biological complexity. *PNAS*, **97**, 9, 4463–4468.

Barrow, J. D. (1994). *The Origin of the Universe*. New York: Basic Books.

Barrow, J. D. (2011). Personal communication.

Bejan, A. (2006). *Advanced Engineering Thermodynamics*, 3rd edn. New York: Wiley.

Bennett, C. H. (1987). Demons, engines and the second law. *Scientific American*, Nov., pp. 108–116.

Bennett, C. H. (1988). Information, dissipation, and the definition of organization. In D. Pines (ed.), *Emerging Syntheses in Science*. Santa Fe: Addison-Wesley.

Bennett, C. H. (1994). Complexity in the Universe. In J. J. Halliwell, J. Perez-Mercader & W. H. Zurek (eds.), *Physical Origins of Time Asymmetry*. Cambridge: Cambridge University Press.

Carroll, S. (2010). *From Eternity to Here: the Quest for the Ultimate Theory of Time*. New York: Dutton, Penguin.

Chaisson, E. J. (2001). *Cosmic Evolution: the Rise of Complexity in Nature.* Cambridge: Harvard University Press.

Chaitin, G. (1987). *Algorithmic Information Theory.* Cambridge: Cambridge University Press.

Crutchfield, J. P. & Young, K. (1989). Inferring statistical complexity. *Phys. Rev. Lett.,* **63,** 105–108.

Davies, P. C. W. (1974). *The Physics of Time Asymmetry.* Berkeley: University California Press.

Davies, P. C. W. (1994). Stirring up trouble. In Zurek, W. H., Perez-Mercader, J., & Halliwell, J. J. (eds.), *Physical Origins of Time Asymmetry.* Cambridge: Cambridge University Press, pp. 119–130.

Deutsch, D. (1997). *The Fabric of Reality.* New York: Penguin, p. 179.

Dopita, M. A., Krauss, L. M., Sutherland, R. S. *et al.* (2011). Re-ionizing the Universe without stars. *Astrophys. Space Sci.,* **335,** 345–352.

Dyson, F. J. (1979). Time without end: physics and biology in an open Universe. *Rev. Mod. Physics,* **51,** 447–460.

Egan, C. & Lineweaver, C. H. (2010). A larger entropy of the Universe. *Astrophysical Journal,* **710,** 1825–1834.

Evans, D. & Searles, D. J. (1994). Equilibrium microstates which generate second law violating steady states. *Physical Review,* E, **50,** 2, 1645–1648.

Frautschi, S. (1988). Entropy in an expanding Universe. In B. H. Weber, D. J. Depew & J. D. Smith (eds.), *Entropy, Information, and Evolution: New Perspectives on Physical and Biological Evolution.* Cambridge, MA: MIT Press, pp. 11–22.

Fullsack, M. (2011). Complexity and its observer: does complexity increase in the course of evolution? Paper presented at the 11th Congress of the Austrian Philosophical Society (OeGP), University of Vienna.

Gell-Mann, M. (1994). *The Quark and the Jaguar: Adventures in the Simple and Complex.* New York: W. H. Freeman.

Gell-Mann, M. (1995). What is complexity? *Complexity,* **1,** no. 1.

Gell-Mann, M. & Lloyd, S. (2003). Effective complexity. In M. Gell-Mann & C. Tsallis (eds.), *Nonextensive Entropy – Interdisciplinary Applications.* USA: Oxford University Press, pp. 387–398.

Hazen, R. M., Papineau, D., Bleeker, W. *et al.* (2008). Mineral evolution. *American Mineralogist,* **93,** 1693–1720.

Hazen, R. M. & Eldredge, N. (2010). Themes and variations in Complex systems. *Elements,* **6,** 43–46.

Jaynes, E. T. (1989). Clearing up mysteries – the original goal. In J. Skilling (eds.), *Maximum Entropy and Bayesian Methods.* Dordrecht: Kluwer Academic Publishing, pp. 1–27.

Kauffman, S. (1995). *At Home in the Universe: the Laws of Complexity*. London: Penguin.

Kleidon, A. (2010). Life, hierarchy and the thermodynamics machinery of planet Earth. *Physics of Life Reviews*, doi:10.1016/j.plrev.2010.10.002.

Kleidon, A. (2012). How does the Earth system generate and maintain thermodynamic disequilibrium and what does it imply for the future of the planet? *Phil. Trans. R. Soc A*, **370**, 1012–1040.

Kleidon, A. & Lorenz, R. D. (2005). *Non-equilibrium Thermodynamics and the Production of Entropy: Life, Earth and Beyond*. Heidelberg: Springer.

Kolb, E. W. & Turner, M. S. (1990). *The Early Universe*. New York: Addison-Wesley.

Krauss, L. & Starkman, G. (2000). Life, the Universe and nothing: life and death in an ever-expanding Universe. *Astrophysical Journal*, **531**, 22–30.

Layzer, D. (1975). The arrow of time. *Scientific American*, **233**, 6, 56–69.

Layzer, D. (1988). Growth of order in the Universe. In B. H. Weber, D. J. Depew and J. D. Smith (eds.), *Entropy, Information, and Evolution: New Perspectives on Physical and Biological Evolution*. Cambridge, MA: MIT Press, pp. 23–39.

Li, M. & Vitanyi, P. M. B. (2008). *An Introduction to Kolmogorov Complexity and Its Applications*. 3rd ed., New York: Springer.

Liddle, A. R. & Lyth, D. H. (2000). *Cosmological Inflation and Large-Scale Structure*. Cambridge: Cambridge University Press.

Lineweaver, C. H. (2001). An estimate of the age distribution of terrestrial planets in the Universe: quantifying metallicity as a selection effect. *Icarus*, **151**, 307–313.

Lineweaver, C.H. (2010). Spreading the power: commentary on life, hierarchy, and the thermodynamic machinery of planet Earth by A. Kleidon. *Phys. Life Rev.*, doi:10.1016/j.plrev.2010.10.004.

Lineweaver, C. H. & Egan, C. (2008). Life, gravity and the second law of thermodynamics. *Physics of Life Reviews*, **5**, 225–242.

Lineweaver, C. H. & Egan, C. (2011). The initial low gravitational entropy of the Universe as the origin of design in nature. In R. Gordon, L. Stillwaggon-Swan & J. Seckbach (eds.), *Origins of Design in Nature*. Dordrecht: Springer, pp. 3–16.

Lloyd, S. (2001). Measures of complexity: a non-exhaustive list. *IEEE Control Systems Magazine*.

Lloyd, S. & Pagels, H. (1988). Complexity as thermodynamic depth. *Annals of Physics*, **188**, 186–213.

Maxwell, J. C. (1888). *Theory of Heat*. London: Longmans, Green and Co.

McShea D. W. (2000). Functional complexity in organisms: parts as proxies. *Biological Philosophy*, **15**, pp. 641–668.

McShea, D. W. & Brandon, R. N. (2010). *Biology's First Law: the Tendency for Diversity and Complexity to Increase in Evolutionary Systems*. Chicago: University of Chicago Press.

Penrose R. (1979). Singularities and time-asymmetry. In Hawking, S. W. & Israel, W. (eds), *General Relativity: an Einstein Centenary Survey*. Cambridge: Cambridge University Press, pp. 581–638.

Penrose R. (2004). The big bang and its thermodynamic legacy. In *Road to Reality: a Complete Guide to the Laws of the Universe*. London: Vintage Books, pp. 686–734. Plot used in Fig. 1, panel c, from A. Thomas (2009), www.ipod.org.uk/reality/reality_arrow_of_time.asp.

Prigogine I. (1978). Time, structure and fluctuations. *Science*, **201**, 777–85.

Quinn, H. R. & Nir, Y. (2008). *The Mystery of the Missing Antimatter*. Princeton: Princeton University Press.

Rubi, J. M. (2008). Does nature break the second law of thermodynamics? (also published as "The long arm of the second law"). *Scientific American*, November.

Sakharov, A. D. (1967). Violation of CP symmetry, C-asymmetry and baryon asymmetry of the Universe. *JETP Letters*, **5**, 24–27.

Schneider, E. D. & Sagan, D. (2006). *Into the Cool: Energy Flow, Thermodynamics, and Life*. Chicago: University of Chicago Press.

Schrödinger, E. (1944). *What is life?* Cambridge: Cambridge University Press.

Smoot, G. F., Bennett, C. L., Kogut, A. *et al.* (1992). Structure in the COBE differential microwave radiometer first-year maps. *Astrophysical Journal*, **396**, L1–5.

Spiegelman, S. (1971). An approach to the experimental analysis of precellular evolution. *Quarterly Reviews of Biophysics*, 4(2&3), 213–253.

Susskind, L. (1995) The world as a hologram. *Journal of Mathematical Physics*, **36**, 6377–6396, arXiv:hep-th/9409089.

Szilard, L. (1929). Uber die Entropie verminderung in einem thermodynamischen System bei Eingriffen intelligenter wesen. *Zeitschrift für Physik*, **53**, 840–856.

Traub, J. F., Wasilkowsu, G. W. & Wozniakowski, H. (1983). *Information, Uncertainty, Complexity*. Reading, MA: Addison-Wesley.

Ulanowicz, R. E. & Hannon, B. M. (1987). Life and the production of entropy. *Proc. Royal Soc. London. Series B, Biological Sciences*, **232**, No. 1267, 181–192.

Weaver, W. (1948). Science and complexity. *American Scientist*, **36**, 536.

Weber, B. H., Depew, D. J., & Smith, J. D. (eds.) (1988). *Entropy, Information, and Evolution: New Perspectives on Physical and Biological Evolution*. Cambridge, MA: MIT Press.

Wicken, J. S. (1988). Thermodynamics, evolution, and emergence: ingredients for a new synthesis. In B. H. Weber, D. J. Depew & J. D. Smith (eds.), *Entropy, Information, and Evolution: New Perspectives on Physical and Biological Evolution*. Cambridge, MA: MIT Press, pp. 139–169.

Wilson E. O. (1992). *The Diversity of Life*. Cambridge: Harvard University Press, p. 9.

Zaikowski, L. & Friedrich, J. (eds.) (2008). *Chemical Evolution across Space and Time: from the Big Bang to Prebiotic Chemistry*. American Chemical Society Symposium Series. USA: Oxford University Press.

Zimdahl, W. & Pavon, D. (2001). Cosmological two-fluid thermodynamics. *General Relativity and Gravitation*, **33**, 5, 791–804, arXiv:astro-ph/0005352v1.

Zurek, W. H. (1989). Thermodynamic cost of computation, algorithmic complexity and the information metric. *Nature*, **341**, 119–124.

4 Using complexity science to search for unity in the natural sciences

Eric J. Chaisson

Nature writ large is a mess. Yet, underlying unities pervade the long and storied, albeit meandering, path from the early universe to civilization on Earth. Evolution is one of those unifiers, incorporating physical, biological, and cultural changes within a broad and inclusive cosmic-evolutionary scenario. Complexity is another such unifier, delineating the growth of structure, function, and diversity within and among galaxies, stars, planets, life, and society throughout natural history. This chapter summarizes a research agenda now underway not only to search for unity in Nature but also, potentially and more fundamentally, to quantify both unceasing evolution and increasing complexity by modeling energy, whose flows through non-equilibrium systems arguably grant opportunities for evolution to create even more complexity.

4.1 COSMIC EVOLUTION

Truth be told, I am a phenomenologist – neither a theorist studying Nature from first principles (I'm not smart enough) nor an experimentalist actually measuring things (although I used to). My current philosophy of approach aims to observe and characterize Nature thermodynamically, seeking to explicate a scientific worldview that chronicles systematically and sequentially the many varied changes that have occurred from the big bang to humankind on Earth. I call that epic worldview cosmic evolution.

Complexity and the Arrow of Time, ed. Charles H. Lineweaver, Paul C. W. Davies and Michael Ruse. Published by Cambridge University Press.
© Cambridge University Press 2013.

A suggested definition: *Cosmic evolution is a grand synthesis of all developmental and generational changes in the assembly and composition of radiation, matter, and life throughout the history of the universe.*

The scientific interdiscipline of cosmic evolution as a general study of change is not new; its essence harks back at least 25 centuries to when the philosopher Heraclitus arguably made the best observation ever while noting that "everything flows... nothing stays." This remarkably simple idea is now confirmed by modern scientific reasoning and much supporting data. I have recently reviewed the status of attempts to undergird the eclectic, integrated scenario of cosmic evolution with quantitative analyses, thereby advancing the topic from subjective colloquy to objective empiricism (Chaisson, 2009a, 2009b).

Academic colleagues often quip that history is "just one damn thing after another", implying that natural history, which goes all the way back in time, comprises myriad and diverse, yet unrelated events. By contrast, I have always regarded natural history expansively and seamlessly as a long and continuous narrative not only incorporating the origin and evolution of a wide spectrum of ordered structures, but also connecting many of them within an overarching framework of understanding. In short, my scientific scholarship firmly roots my work in empirical research, mines data from a wealth of observations across all of space and time, and portrays natural history as an intellectually powerful story that unifies much of what is known about Nature.

Although guiding changes within and among complex systems, evolution itself need not be a complex process. Nor does evolution, as an erratic, rambling activity that is unceasing, uncaring, and unpredictable likely pertain only to life forms. Cosmic evolution extends the central idea of evolution – ascent with modification, generally considered – to embrace all structured systems. And by merging physical, biological, and cultural evolution into a single, intensive paradigm based on everlasting change, cosmic evolution evokes a Platonic ideal that the changing, shifting world of natural phenomena and realistic objects masks a deeper, underlying reality of unchanging forms and processes, and that it is these alone that grant true knowledge.

4.2 ENERGY RATE DENSITY

All complex systems – whether living or not – are open, organized, non-equilibrated structures that acquire, store, and express energy. This chapter's single goal reiterates and amplifies a previously proposed hypothesis (Chaisson, 2001) that specific energy flow reifies a complexity metric and potential evolutionary driver for all constructive events from the origin of the universe to humans on Earth, as well as for future evolutionary events yet to occur. Energy does seem to be a common currency among such ordered structures; energy flow may well be the most unifying process in science, helping to provide a cogent explanation for the onset, existence, and complexification of a whole array of systems – notably, how they emerge, mature, and terminate during individual lifetimes as well as across multiple generations.

Energy is not likely the only useful metric to measure complexity in complex, evolving systems. Nor do I mean to be critical of alternative schemes, such as information content or entropy production; the literature is replete with controversial claims for such measures, many of them asserted with dogmatic confidence. I have earlier published brief critiques that these and related alternatives are unhelpful for general complexity metrics, their use often narrow, abstract, qualitative, and equivocal (Chaisson, 2001). By contrast, I have embraced the practical concept of energy largely because I can define it, measure it, and clearly express its units. I have furthermore endeavored to quantify this decidedly thermodynamic term in a reliable and consistent manner for a full spectrum of organized systems from spiral galaxies and fusing stars to buzzing bees and redwood trees, indeed to sentient humans and our technological society.

The chosen metric, however, can be neither energy alone, nor even merely energy flow. Life on Earth is surely more complex than any star or galaxy, yet the latter engage vastly more energy than anything now alive on our planet. Accordingly, I have sought to normalize energy flows in complex systems by their inherent mass, thereby enabling more uniform analysis while allowing effective comparison between and among virtually every kind of system encountered in Nature. This, then, has been and continues to be my working hypothesis: mass-normalized energy flow, termed energy rate density and denoted by Φ_m, is potentially the most universal process capable of building structures, evolving systems, and creating complexity throughout the universe.

A suggested definition: *Energy rate density (also termed power density) is the amount of energy flowing through a system per unit time and per unit mass.*

For consistency in this research program's calculations, I have used total energy flowing through the bulk of open systems since all incoming energy passing through such systems is eventually dissipated regardless of the efficiency with which systems utilize energy. A more refined analysis might benefit from using either the physicist's "free energy" or the chemist's "enthalpy", although for well-organized systems internal energy and free energy are nearly the same, and in any case the general results of this study would not likely change much given the ten-order-of-magnitude trend in energy rate density from galaxies to society. Several other recent works have also employed the concept of energy rate density, albeit in more limited venues (e.g., Spier, 2011; Neubauer, 2011).

Figure 4.1 summarizes much recent research on this subject, depicting how physical, biological, and cultural evolution over ~14 Gy has transformed homogeneous, primordial matter into increasingly intricate systems (Chaisson, 2011a, 2011b). The many graphs show the rise in values of Φ_m computed for selected systems extant in Nature and of known scientific age. (For specific power units of W/kg, divide by 10^4.) Values given are typical for the general category to which each system belongs, yet as in any simple, unifying explication of an imperfect universe – especially one like cosmic

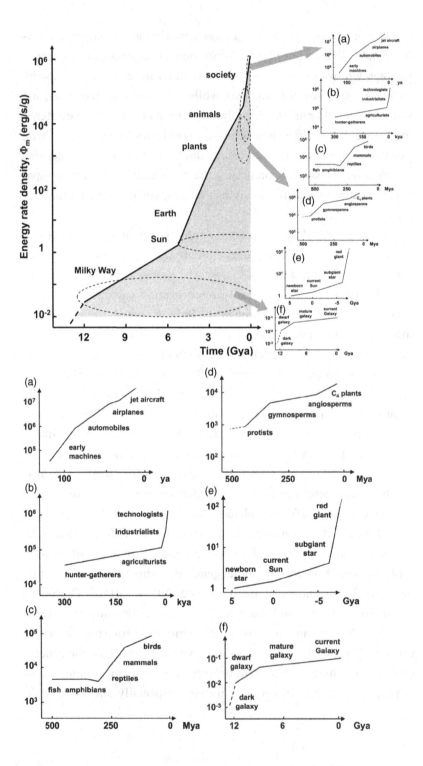

evolution that aspires to address all of Nature – there are moderate variations. And it is likely that from those variations arose the great diversity among complex, evolving systems everywhere.

Better metrics than energy rate density may well describe each of the individual systems within the realms of physical, biological, and cultural evolution that combine to create the greater whole of cosmic evolution, but no other single metric seems capable of uniformly describing them all. The significance of plotting "on the same page" a single quantity for such a wide range of systems observed in Nature should not be overlooked. I am unaware of any other sole quantity that can characterize so extensively a principal system dynamic over >20 orders of magnitude in spatial dimension and nearly as many in time.

What seems inherently attractive is that energy flow as a universal process helps suppress entropy within increasingly ordered, localized systems evolving amidst increasingly disordered, surrounding environments, indeed a process that arguably governed the emergence and maturity of our Galaxy, our star, our planet, and ourselves. All accords with the second law of thermodynamics; no violations or circumventions of Nature's most cherished law are evident. If

FIGURE 4.1 These many graphs show changing values of energy rate density, Φ_m, for myriad systems observed throughout Nature. The main graph at the top left traces Φ_m for a variety of open, organized, non-equilibrium systems extending from the big bang to humankind. Plotted semi-logarithmically at the time of each system's origin, Φ_m displays a clear increase during the ~14 Gy history of the universe. The shaded area includes a huge ensemble of changing Φ_m values as individual systems evolved and complexified. The dashed ovals outline the range in Φ_m and time bracketing each of the physical, biological, and cultural systems graphed at top right (each magnified at bottom). Rationale for the main plot on the left can be found in Chaisson (2001); data for all the plots on the top right are from Chaisson (2011a, 2011b). Exceptions, outliers, "black swans", or whatever one wants to call those data points that inevitably deviate from the norm, are occasionally evident. The Φ_m values and historical dates plotted here are estimates, each with ranges and uncertainties; yet it is not their absolute magnitudes and specific quantities that matter as much as their overall trend with the march of time.

correct, energy itself is a central mechanism of change – a central feature of evolution. And energy rate density is an unambiguous, weighted measure of energy flow enabling us to gauge all complex systems in like manner, as well as to examine how over the course of time some systems were able to command energy and survive, while others apparently could not and did not.

4.3 COMPLEXITY QUANTIFIED

Cosmic evolution is not a theory of everything, nor even necessarily a universal theory of evolution; it is, rather, a collection of evolutionary phases – from rudimentary alteration of physical systems, to Darwinian modification of life forms, to Lamarckian reshaping of cultured society – all consistently and fundamentally characterized, at least in part, by mass-normalized energy flow. All complex systems, samples of which are diagnosed below (see, Chaisson, 2011a, 2011b), interact with their environments as matter and energy flow in while wastes flow out, adapt to changing circumstances, and resemble metabolisms at work on many scales. These findings strengthen the time-honored idea that elegantly simple processes underlie the tangled complexity of our richly endowed universe.

A suggested definition: *Complexity is a state of intricacy, complication, variety or involvement, as in the interconnected parts of a system – a quality of having many different, interacting components.*

Physical evolution

Stars and galaxies among physical systems generally have energy rate densities that are among the lowest of known organized systems. The latter, including those of dwarf, normal, and active galaxies, display $\Phi_m = 0.01$–50 erg/s/g, each type showing clear temporal trends in rising values of Φ_m while clustering hierarchically, as herewith computed for our Milky Way Galaxy:

- from protogalactic blobs >12 Gya ($\Phi_m \approx 10^{-3}$ erg/s/g),
- to widespread dwarf galaxies ($\sim 10^{-2}$),
- to mature, normal status ~ 10 Gya (~ 0.05),
- to our Galaxy's current state (~ 0.1).

Although of lesser complexity and longer duration, the Milky Way is nearly as adaptive and metabolic as any life form – transacting energy while forming new stars, cannibalizing dwarf galaxies, and dissolving older components. Stars, too, adjust their states while evolving during one or more generations, their Φ_m values rising while they complexify with time. Stellar interiors undergo cycles of nuclear fusion that foster greater thermal and chemical gradients, resulting in increasingly differentiated layers of heavy elements within highly evolved stars. Stellar size, color, brightness, and composition all change while slowly altering the structure of every star, including the Sun, which will eventually be selected out of the population of neighboring stars:

- from early protostar ~5 Gya ($\Phi_m \approx 1$ erg/s/g),
- to the main-sequence Sun currently (~2),
- to subgiant status ~6 Gy in the future (~4),
- to aged red giant near termination (~10^2).

At least as regards energy flow, material resources, and structural integrity while experiencing change, adaptation, and selection, stars have much in common with life. This is not to say that stars are alive, nor that stars evolve in the strict and limited biological sense; most researchers would agree that stars and galaxies develop – as evidenced by systematically rising Φ_m values.

Biological evolution

In turn, plants and animals among biological systems regularly exhibit intermediate values of $\Phi_m = 10^3 - 10^5$ erg/s/g. Life does seem to operate optimally within certain limits of temperature, pressure, salinity, etc., and not surprisingly also has an optimal range of normalized energy flow. For plant life on Earth, energy rate densities are much higher than those for galaxies, stars, and planets, as perhaps best illustrated by the evolution of the most dominant process in Earth's biosphere – photosynthesis:

- from microscopic protists >470 Mya ($\Phi_m \approx 10^3$ erg/s/g),
- to gymnosperms ~350 Mya (~5×10^3),
- to angiosperms ~125 Mya (~7×10^3),
- to highly efficient C_4 plants ~30 Mya (~10^4).

Onward across the bush of life (or the arrow of time) – cells, tissues, organs, organisms – much the same metric holds for animals while evolving and complexifying. For adult bodies (much as for brains, which have an order of magnitude larger Φ_m), the temporal trend of rising Φ_m continues:

- from fish and amphibians 370–500 Mya ($\Phi_m \approx 4 \times 10^3$),
- to cold-blooded reptiles ~320 Mya (~3×10^3),
- to warm-blooded mammals ~200 Mya (~4×10^4),
- to birds in flight ~125 Mya (~9×10^4).

Here, system functionality and genetic inheritance, two factors above and beyond mere system structure, help to enhance complexity among animate systems that are clearly living compared to inanimate systems that are clearly not. In either case, energy is fuel for change, apparently (and partly) selecting systems able to utilize increased power densities, while driving others to destruction and extinction – all in accord with neo-Darwinism's widely accepted modern synthesis.

A suggested definition: *Life is an open, coherent, spacetime structure kept far from thermodynamic equilibrium by a flow of energy through it – a carbon-based system operating in a water-based medium, with higher forms metabolizing oxygen.*

Cultural evolution

Among cultural systems, advances in technology compare to those of society itself, each of them energy-rich and with $\Phi_m \geq 10^5$ erg/s/g – hence plausibly the most complex systems known. Social progress can be tracked, again in terms of energy consumption, for a variety of human-related cultural advances among our human ancestors:

- from hunter-gatherers ~300 kya ($\Phi_m \approx 4 \times 10^4$ erg/s/g),
- to agriculturists ~10 kya (~10^5),
- to industrialists ~200 ya (~5×10^5),
- to technologists of today (~2×10^6).

Machines, too, and not just computers, but also ordinary motors and engines that typified the fast-paced economy of the twentieth century, can be cast in evolutionary terms – though here the mechanism is less Darwinian than Lamarckian, with the latter's emphasis on accumulation of acquired traits. Either way, energy remains a driver, and with rapidly accelerating pace:

- from primitive machines ~150 ya ($\Phi_m \approx 10^5$ erg/s/g),
- to the invention of automobiles of ~100 ya ($\sim 10^6$),
- to the development of airplanes ~50 ya ($\sim 10^7$),
- to computerized jet aircraft of today ($\sim 5 \times 10^7$).

The road to our present technological society was doubtlessly built with increased energy density used, or per capita energy expended. Increasingly sophisticated technical gadgets, under the Lamarckian pressure of dealer competition and customer selection, do in fact show increases in Φ_m values with product improvement over the years. The cultural evolution of many silicon-based devices now central to our global economy can likewise be traced and their rising Φ_m values computed, the two – evolution and complexity – paralleling each other once again.

4.4 SUMMARY

Complexity science is less empirical and encompassing than many practitioners admit. Traditionally, this subject probes diverse collections of distinct topics, such as cells, ants, economies, and networks, while often appealing to information theory to decipher general principles of mostly biological and social systems that display emergent and adaptive qualities. Such efforts have garnered limited success and an unusual amount of controversy for such a promising new field. Although yielding insightful properties of systems unlikely to be understood by reductionism alone, the real promise of complexity science remains as elusive as when it first arose a generation ago.

This chapter proffers a different strategy. It goes beyond mere words, indeed beyond specialized disciplines, to explore widely, deeply, and phenomenologically a process that might characterize complexity quantitatively across many scientific domains. I have assessed a great array of systems, sought commonalities among them all, and examined a single, uniform metric that arguably quantifies the observed rise of complexity among Nature's many varied systems. The result is an expansive evolutionary scenario not only spanning the known history of time to date but also revealing strong similarities among systems as disparate as stars, life, and society.

Cosmic evolution is more than a subjective, qualitative narration of one unrelated event after another. This inclusive scientific worldview constitutes an objective, quantitative approach toward deciphering much of what comprises organized, material Nature. It addresses the coupled topics of system change and complexity – the temporal advance of the former having contributed to spatial growth of the latter, yet the latter feeding back to make the former increasingly productive. It demonstrates that the basic differences, both within and among many varied complex systems, are of degree, not of kind. And it suggests that optimal ranges of energy rate density grant opportunities for the evolution of complexity; those systems able to adjust, adapt, or otherwise take advantage of such energy flows survive and prosper, while other systems adversely affected by too much or too little energy are non-randomly eliminated. All things considered, I conclude the following:

- Evolution is a universal phenomenon; including changes in physical, biological, and cultural systems, evolution is a unifying principle throughout natural science.
- Energy is a common currency; energy rate density (Φ_m) generally correlates with system complexity and may drive, at least in part, the process of evolution itself.
- Selection and adaptation are ubiquitous in Nature; the emergence, maintenance, and fate of all complex systems are often determined, again partly, by their ability to utilize energy.

Physicists tend to notice large trends and general patterns in Nature, often seeking grand unifications or at least global explanations based on few and simple tenets. Biologists, by contrast, concentrate on minute details and intricate mechanisms, often noting quite rightly rare abnormalities in the sweeping generalities. Such dual attitudes perhaps signal the true value of this coarse-grained, phenomenological approach, for only when the devilish details are reconciled with the bigger picture will we be able to call it a "complexity science" that synthesizes both for coherent understanding of ourselves, our world, and our universe.

REFERENCES

Chaisson, E. J. (2001). *Cosmic Evolution: the Rise of Complexity in Nature*. Cambridge & London: Harvard University Press.

Chaisson, E. J. (2009a). Cosmic evolution – state of the science. In S. Dick and M. Lupisella (eds.). *Cosmos & Culture*. Washington: NASA Press.

Chaisson, E. J. (2009b). Exobiology and complexity. In R. Meyers (ed.). *Encyclopedia of Complexity and Systems Science*. Berlin: Springer.

Chaisson, E. J. (2011a). Energy rate density as a complexity metric and evolutionary driver. *Complexity*, **16**, 27–40; DOI: 10.1002/cplx.20323.

Chaisson, E. J. (2011b). Energy rate density II: probing further a new complexity metric. *Complexity*, **17**, 44–63; DOI: 10.1002/cplx.20373.

Neubauer, R. L. (2011). *Evolution and the Emergent Self: the Rise of Complexity and Behavioral Versatility in Nature*. New York: Columbia University Press.

Spier, F. (2011). *Big History and the Future of Humanity*. London: Wiley-Blackwell.

5   On the spontaneous generation of complexity in the universe

Seth Lloyd

A glance out the window confirms that the universe is complex. By day, intricate weather patterns chase the Sun along its path. At night, galaxies, stars, and planets wheel across the sky. Rabbits hop across the lawn, pursued by coyotes. Trucks rumble along the highway. Turning one's gaze inside the room confirms the diagnosis. People chat and argue. Children grow. Food cooks on the stove. Spam accumulates in the computer inbox.

How and why did all this complexity come about? The answers that we currently possess are largely qualitative and historical. The big bang happened, gravitational instability made matter clump together, stars started to shine, planets formed, life began, humans showed up, societies formed, all hell broke loose. We know that the universe is complex, and we know something of the sequence in which more and more complex systems developed. It would be good, however, to know WHY the universe is complex. Is there some intrinsic drive to the creation of complexity in matter and energy? Are there other universes, and if so, are they more or less complex than ours? Will this generation of complexity go on for ever?

To answer such questions in a precise, scientific (i.e., falsifiable) way is not so easy. First of all, how can we measure complexity? Most of us regard complexity in the way that Justice Potter Stewart regarded pornography – we may not be able to define it, but we know it when we see it. If we want to identify conditions under which complexity spontaneously arises out of the combination of chance and causality, knowing complexity when we see it is insufficient. To answer the

Complexity and the Arrow of Time, ed. Charles H. Lineweaver, Paul C. W. Davies and Michael Ruse. Published by Cambridge University Press.
© Cambridge University Press 2013.

question why complexity arises, we need some more precise notion of what complexity is.

The second problem that faces us in identifying sources of complexity is ignorance of how complexity arose in the first place. We know more about the origins of the universe itself than we know about the origins of life on Earth. For better or worse, the universe seems to have begun in a particularly simple, non-complex state. Extrapolating current cosmological observations to the time of the big bang shows that the early universe was uniform, flat, and featureless. The early universe was simple. Moverover, the known laws of physics are simple – they can be written down on the back of a tee-shirt. How can such initial simplicity transformed by simple laws give rise to all subsequent complexity? To analyze how complexity comes into existence, we need to know more about the dynamics of both simple and complex systems.

The primary purpose of this chapter is to report on what we know and what we don't know about the spontaneous generation of complexity in the universe. Our ignorance is vast, of course, but we also possess a surprising amount of knowledge about how to define complexity, and about the types of dynamics that spontaneously give rise to complexity. This chapter reviews the state of the art and science of what complexity is and how it arises. By answering the questions What and How, we may get closer to answering the question Why.

After marshalling relevant facts from physics, cosmology, and computer science, this chapter presents a particular mechanism for generating complexity. I argue that the universe generates complexity because the universe is – in a mathematically precise sense – a quantum computer (Lloyd, 1993). The intuition behind this mathematical equivalence is straightforward: the laws of physics in our universe allow the construction of both classical and quantum computers (Nielsen & Chuang, 2000), and quantum computers in turn can efficiently simulate the laws of physics (Lloyd, 1996). Two systems that can efficiently simulate each other possess the same computational abilities. One can then prove (Lloyd, 1997) that a quantum computer, programmed to perform all possible programs in quantum parallel, generates all computable structures, including structures of arbitrarily high complexity and intricacy. (Quantum parallelism is an effect in which different programs are evaluated in different components of the wave function in quantum superposition.) This proof implies that our universe, programmed by quantum fluctuations and evolving by computationally universal laws, necessarily transforms energy and information to create complex and intricate structures, including the ones that we see when we look out of the window (Lloyd, 2006).

5.1 MARSHALLING THE FACTS

This chapter provides a comprehensive review of the quantum computational argument for the generation of complexity. The argument is based on three kinds of fact: (1) empirical facts about the universe derived from the combination of observation and theory, (2) theoretical physics and chemistry, and (3) mathematics. In practice, there is no strict dividing line between the first two forms: observation is always informed by theory, and even "pure" physical theory is informed by observation. Mathematical results, by contrast, are purely theoretical.

(1) Empirical/theoretical observations
  (1.1) The observed universe is complex and exhibits structure at all scales, ranging from the most microscopic distances accessible to, e.g., particle accelerators, to distances approaching the size of the observationally accessible universe.
  (1.2) Detailed astronomical observations combined with the known laws of physics for elementary particles and gravitation suggest that the universe is infinite in spatial extent (Liddle & Lyth, 2000; Mukhanov, 2005; Weinberg, 2008).
  (1.3) Although the universe as a whole may be infinite in time and space, the part of the universe that can be accessed by our direct observations is finite in spatial and temporal extent. The boundary of this accessible part of the universe is called the particle horizon. The behavior of matter outside the horizon can only be

obtained by extrapolation from observations inside the horizon combined with physical theory.

(1.4) The galaxies and clusters of galaxies within the horizon are moving apart from each other – the observable part of the universe is expanding. The further away a galaxy is, the faster it is moving away from us. Extrapolating backwards in time leads to an initial explosive event – the big bang – 13.75 ± 0.13 billion years ago.

(1.5) Over the past few decades, observations of distant galaxies strongly suggest that the expansion of the universe is accelerating. Even though the gravitational attraction between galaxies tends to slow the expansion of the observed universe, a mysterious form of energy called dark energy is apparently causing the universe to expand more and more rapidly. The dark energy gives rise to a term in Einstein's equations called the cosmological constant, although whether or not the dark energy truly possesses constant density is not known. The degree of acceleration of the expansion is very small, but it will eventually dominate the observed structure of the universe, moving our Galaxy out of causal contact with all other galaxies except our immediate neighbors.

(1.6) At its most microscopic scales, the dynamics of the universe is described by the laws of quantum mechanics. Quantum mechanics governs the behavior of elementary particles – it determines how those elementary particles can assemble themselves into atoms – it determines how atoms assemble themselves into molecules – it governs the dynamics of chemical reactions (Nielsen & Chuang, 2000; Messiah, 1999; Peres, 2002).

(2) Theoretical physics

(2.1) General relativity is the physical theory that governs physics at very large scales. Newtonian gravity is often a good approximation to general relativity when the timescale of gravitational dynamics is slower than or even comparable to the time it takes light to pass between the gravitationally interacting bodies (Liddle & Lyth, 2000; Mukhanov, 2005; Weinberg, 2008).

As just noted, quantum mechanics governs physics at smaller scales (Messiah, 1999; Peres, 2002). Classical mechanics is often a good approximation to physical behavior at intermediate length and timescales (e.g. one meter, one second). Quantum mechanics is intrinsically probabilistic, while classical mechanics is deterministic. So a classical mechanical simulation of a quantum system frequently requires a source of randomness.

One of the embarrassments of theoretical physicists is their long-standing inability to construct a theory of quantum gravity that both is theoretically self-consistent and can be used to make firm empirical predictions/retrodictions. Existing theories of quantum gravity such as loop quantum gravity and string theory fall short on both accounts.

(2.2) Despite the lack of a good, widely-accepted theory of quantum gravity, some aspects of the behavior of quantum fields on curved-spacetime are well established and trusted (Birrell & Davies, 1982). Such partial theories of quantum mechanics and general relativity allow physicists to extrapolate backwards before the big bang. Currently, the most widely accepted pre-big bang theory is called inflation (Guth, 1981; Starobinsky, 1982; Linde, 1986a, 1986b; Vilenkin, 1994; Aguirre et al., 2010). In inflation, the large amount of energy in the quantum vacuum causes the scale factor of the universe to double in size again and again on very short timescales (perhaps as short as the Planck time, ca. 5.4×10^{-44} seconds). As inflation continues, the wave function of the universe moves down a potential landscape, accumulating energy and releasing it when it reaches a potential minimum. This released energy thermalizes yielding a hot, dense, flat, expanding universe – the initial state of the hot big bang.

The region undergoing inflation is infinite: our observed portion of the universe is just part of this infinite whole. Different sub-regions in this infinite expanse are governed by the same laws of physics as in our region, but start from a variety of different initial conditions. Tegmark (2007; Aguirre et al., 2010) calls this situation a "type I multiverse": if we call the finite area within our particle horizon "the observable universe", the type I multiverse contains many such universes.

Inflation is widely accepted because some inflationary models correctly retrodict observed features of our part of the universe, including its approximate spatial homogeneity, isotropy, flatness, and the fact that spatially distant parts of the universe have apparently been in causal contact in the distant past. Since it is the most well-established and observationally confirmed cosmological theory, our discussion of the generation of complexity in the universe will focus on the type I multiverse – i.e., the observable universe and region beyond the horizon.

(2.3) Extrapolating the idea of inflation beyond regions for which we have observational evidence leads to a model known as "eternal inflation" (Linde, 1986a, 1986b; Vilenkin, 1994). In this model, our (infinite) post-inflation sector of the universe is connected to an infinite number of other such sectors, which can be governed by different laws of physics. The different sectors or "bubbles" are still presumably governed by quantum mechanics and general relativity, but they may possess different symmetries that govern the interactions between elementary particles, different numbers of spatial dimensions, etc. The different post-inflation bubbles are separated by regions with high inflation, which – if sufficiently large to begin with – prevent communication from bubble to bubble.

Tegmark (2007) calls this collection of bubbles the "type II" multiverse. The existence of the type II multiverse is less well established than the type I, for which there is direct observational evidence. Nonetheless, extrapolating various theories of eternal inflation typically leads to a type II multiverse picture.

(2.4) The laws of quantum mechanics together with general relativity imply that systems with bounded energy and bounded volume can have only a finite number of distinguishable states (Lloyd, 2000, 2002). This number can be very large – the matter and energy contained within our particle horizon have about $2^{10^{92}}$ possible states. Nonetheless this number is finite. Accordingly, a space-like line drawn from Earth in any direction will eventually pass through the particle horizon into a region with different configurations for the matter. After passing through on the order of $2^{10^{92}}$ volumes the same size as the observationally accessible part of the universe, this line will reach a region identical to our own.

Similarly, in eternal inflation, there are only a finite number of possible sets of physical laws and initial conditions that can govern different bubbles. (String theory suggests a lower number of ca. 10^{500} different sets of laws, see Susskind, 2007.) Since the eternally inflating universe contains an infinite number of such bubbles, each bubble is replicated an infinite number of times.

(2.5) Because there may exist other regions of the eternally inflating universe – other bubbles – with entirely different laws of physics, I will frequently refer to the laws of physics within the observationally accessible part of the universe – which lies inside our bubble – as "our" laws of physics. Our laws of physics clearly support galaxies, stars, and planets, which in turn support highly complex chemical, biological, and social structures. Whether or not other bubbles contain such complexity is a question that will be discussed in detail below. Our theory of the generation of complexity will be applicable not just to our observed universe, but to bubbles governed by laws of physics that are different from our own.

(2.6) The laws of physics in the observed part of the universe – our laws – support computation.

(3) Mathematics

This third section of facts relevant to the generation of complexity in the universe consists of mathematics. Mathematical facts possess a different ontological status from observational/theoretical physical facts. A particular set of observational/theoretical physical facts might only hold in one particular type of bubble, for example. By contrast, many if not most mathematicians regard mathematical facts as possessing a kind of "Platonic" existence separate from any actual physical instantiation as part of the universe. In the following list, I will be careful to distinguish between mathematical facts in the abstract and the physical instantiation of mathematics. This distinction is important for our theory as the primary mathematical facts that we will use are taken from computer science. Pretty much everyone is familiar with actual electronic computers as physical systems within our part of the universe. In addition to their physical manifestions in our world, however, mathematical abstractions of computers (digital, analog, quantum) inhabit the Platonic realm of pure mathematics.

(3.1) The abstract concept of a digital computer is given by a Turing machine (Turing, 1937; Papadimitriou & Lewis, 1982). A Turing machine consists of a tape, divided into squares, and a head located at one of the squares. Each square of the tape contains one of a finite set of symbols. The head can be in one of a finite set of states. The operation of a Turing machine proceeds in discrete time steps. The head reads the symbol on its current square, updates both that symbol and its own internal state, and moves one step to the left or the right along the tape. The updating and move are deterministic functions of the symbol on the current square and the current state of the head.

The tape is finite, but "infinitely extensible": if the head reaches the end of the tape without completing the computation and halting, a new square containing a special symbol (e.g, "blank") is automatically appended.

(3.2) Turing machines perform a particular type of digital computation. Some Turing machines are universal: they can be programmed to simulate any other Turing machine.

(3.3) Two abstract models of computation, A and B, are considered to be computationally equivalent if model A can efficiently simulate model B and vice versa. Here, "efficiently" means using some bounded amount of resources to perform the simulation: different types of bounds lead to different notions of computational equivalence. This chapter will use the notion of polynomial equivalence. If A's computation performs n elementary logical operations on m bits, then B's simulation is efficient if it can be performed in $p(m, n)$, where $p(m, n)$ is a polynomial in m, n. Universal Turing machines represent a powerful abstract notion of computation. If we let A be the Turing machine model of computation and B be the mathematical abstraction of the type of computation performed by familiar electronic digital computers, then it is not hard to show that A and B are computationally equivalent: Turing machines can simulate efficiently the logical steps carried out by electronic digital computers and vice versa. Whether or not our electronic digital computers are actually capable of physically performing any computation performable in the abstract on a Turing machine is a practical and cosmological question: it hinges on whether we (and our descendants) will always be able to supply our computers with more memory space as required.

(3.4) The mathematical definition of computational equivalence implies that many systems that may not at first glance appear to be computers are actually computing, and are capable of universal computation. For example, simple digital models of colliding molecules in a gas ("lattice gases") are computationally equivalent to universal Turing machines and to digital computers (Margolus, 1984). A wide variety of physically inspired digital systems, e.g., cellular automata, are capable of universal computation (Wolfram, 1986).

(3.5) A quantum computer is a digital computer that operates according to the laws of quantum mechanics (Nielsen & Chuang, 2000). The bits in a quantum computer are quantum bits or "qubits". Where a classical bit can be either in the logical state 0 or in the logical state 1, a quantum bit can be in coherent quantum superpositions $\alpha|0\rangle + \beta|1\rangle$ of the quantum logical state $|0\rangle$ and the quantum logical state $|1\rangle$. Quantum computers operate by putting their qubits in superpositions and by performing superpositions of computations. Quantum computers are at least as powerful as classical digital computers: a quantum computer can simulate a classical digital computer with polynomial efficiency, but there is no known algorithm that allows a classical digital computer to simulate a quantum computer efficiently.

(3.6) Quantum computers can simulate the known laws of physics efficiently (Lloyd, 1996). More precisely, they can simulate efficiently any quantum system that has bounded local energy and that evolves according to local interactions. Although we don't have a good theory of quantum gravity, if the correct theory is bounded and local, then it can be efficiently simulated on a quantum computer.

(3.7) A wide variety of quantum dynamics supports quantum computation. Essentially any quantum dynamics that supports (a) the propagation of quanta from one place to another, and (b) nontrivial interactions between quanta, supports universal quantum computation (Lloyd, 1995). In particular, the Standard Model of elementary particles governs the propagation and interaction of quanta in our observationally accessible sector of the universe. Experimental demonstrations of prototype quantum computers (Nielsen & Chuang, 2000) show that the Standard Model supports quantum computation, as do various "sub-dynamics" of the standard model, such as quantum electrodynamics. The mathematical requirements for a physical theory to support quantum computation are simple: (1) the theory must be quantum mechanical; (2) it must allow information to propagate from place to place; and (3) it must support non-trivial interactions (Lloyd, 1995). The theory of quantum error correction shows that quantum computers can accurately perform arbitrarily long quantum computations in principle. In practice, current quantum computing technologies are difficult to scale up to large size. For the purposes of analyzing complexity generation, however, what is important is that quantum computation and locally finite quantum theories can be computationally equivalent in the pure mathematical sense that they can each support or simulate each other. Because this is a central point of the paper, I state it explicitly.

(3.8) Quantum computers can simulate the Standard Model efficiently, in principle (Lloyd, 1996), and the dynamics of the Standard Model support quantum computation. The dynamics of the Standard Model and quantum computation are computationally equivalent in the mathematical sense that each can be programmed to simulate the other.

Although we do not have a good predictive theory of quantum gravity, if quantum gravity is locally finite and evolves according to local interactions, then quantum gravity, too, can be efficiently simulated by a quantum computer.

Two systems can be computationally equivalent, and yet look very different. A lattice gas does not in any way resemble a smart phone – yet they both have the same computational power.

We need just a few more mathematical facts before assembling our theory of complexity generation in the universe. These facts have to do with the mathematical characterization of complexity. First, define algorithmic information content (Solomonoff, 1964; Chaitin, 1987; Kolmogorov, 1965; Li & Vitanyi, 2008).

(3.9) The algorithmic information content/algorithmic complexity/ Kolmogorov complexity $K_U(b)$ of a finite bit string b is the length in bits of the shortest program $p_U(b)$ that makes the universal Turing machine U produce b (and only b) as output. That is, algorithmic information content is defined in terms of the particular computer language, such as Java, in which the program is written.

(3.10) Because any program for the universal Turing machine U can be converted into a program for another universal Turing machine U' by a translation program of length (in bits) $c_{U'U}$, the algorithmic information content depends on U only up to a constant: $K_U - c_{UU'} \leq K_{U'} \leq K_U + c_{U'U}$. Accordingly, from now on we will drop the U in $K_U(b)$ unless it is explicitly required.

(There is an important subtlety here. One can always construct some Turing machine with respect to which some very simple constructions, such as "$1 + 1 = 2$", take an agonizingly long program to derive. By contrast, ordinary computer languages such as Java require only brief programs to express conventional mathematical formalism. When estimating algorithmic complexity, we will assume that brief mathematical formulae have a succinct expression.)

Algorithmic information content is low for simple, highly ordered strings, like $00 \ldots 0$, where the 0 is repeated a billion times. The shortest program to produce a billion 0s is at least as short as the one-liner, "Print 0 one billion times". The known laws of physics have low algorithmic information content – they can be printed on the back of a tee-shirt. Similarly, observations of the cosmic microwave background, combined with the inflationary universe scenario, suggest that the state of the universe in the early stages of inflation – the vacuum – had low algorithmic information content. Indeed, written in second quantized form, the vacuum is the state $|00 \ldots 0\rangle$, where each 0 indicates that there are zero particles in the corresponding mode: the vacuum is the physical analogue to the bit string $00 \ldots 0$.

Conversely, long random bit strings such as one constructed by flipping a coin a billion times have algorithmic information content approximately equal to their length. If the bit string resulting from the random process is $00110100 \ldots 01$, for example, then the shortest program to give this string as output is no longer than the program, "Print $00110100 \ldots 01$", which is slightly more than a billion bits long. The great majority of billion bit strings have shortest programs at most slightly fewer than a billion bits long. In particular, counting the number of possible programs shows that at most a fraction 2^{-m} of bit strings can have a shortest program that is m bits shorter than their length.

Intricate strings with many patterns and hidden order, such as the first billion bits of the binary representation of π, for example, or a digital representation of a virus or bacterium, tend to have intermediate algorithmic information content. This is why algorithmic information content on its own is not a good measure of complexity – it is more a measure of randomness. A mathematical definition of complexity that corresponds more closely to our intuitive ideas of complexity is Bennett's logical depth (Bennett, 1985, 1990).

(3.11) The logical depth of a bit string b is the number of elementary logical operations – i.e., the temporal computational complexity – performed by a Turing machine as it computes b from its shortest program. (If a slightly longer program takes much less time, then Bennett suggests defining logical depth in terms of the less onerous computation.)

A complementary measure of complexity is Lloyd and Pagels's (1988) thermodynamic depth.

(3.12) The thermodynamic depth of a bit string b is the amount of memory space – i.e., the spatial computational complexity – required to produce b from its shortest program.

Logical and thermodynamic depth are measures of the amount of effort required to produce a bit string from a brief description. Like algorithmic information content, they are Turing machine independent up to a multiplicative factor measuring the extra number of ops/memory space required to make Turing machine U simulate Turing machine U'.

This ends the review of facts pertinent to analyzing how and why the universe produces complexity. The universe is large – potentially infinite – and arose from an algorithmically simple initial state by algorithmically simple dynamics. The universe is governed at bottom by the laws of quantum mechanics, and at large scales by general relativity. The quantum mechanical physical dynamics of the universe has the same computational power as quantum computation. The theory of computation can be used to construct mathematically well-defined algorithmic measures of information, and of complexity. It will turn out not to be crucial whether we use exactly these definitions of complexity. What is important is that there exist mathematically well-defined measures of complexity at all. The existence of such measures will now allow us to prove that, starting from simple initial conditions and evolving according to simple laws, the universe necessarily produces complex structures.

5.2 THE DUAL PHYSICAL/COMPUTATIONAL NATURE OF THE UNIVERSE

The key ingredient for the production of complexity in the universe is the mathematical fact – discussed above – that the laws of physics are capable of supporting quantum computation, and that quantum computation allows the efficient simulation of the laws of physics. That fact implies that quantum computers and physical laws in the universe possess the same computational power. That is, the universe is mathematically equivalent to a quantum computer in terms of its ability to process information and to create complex and intricate structures. It also means that when we describe the evolution of the universe in terms of physical quantities such as energy and entropy, we can also present an equivalent but complementary description in terms of computational quantities such as logical operations ("ops") and bits. Because the dual computational/physical nature of the universe plays an important part in the generation of complexity, we dwell on it in more detail here.

The relationship between information and entropy has been known since the nineteenth century (Cover & Thomas, 1991). Consider a physical system with different states labelled by j, where the jth state has probability p_j. For example, j could label the different configurations for the positions and velocities of molecules in a gas. Maxwell and Boltzmann defined entropy to be $S = -k_B \sum_j p_j \ln p_j$, where $k_B = 1.38 \times 10^{-23}$ joule/Kelvin is Boltzmann's constant. Comparing this formula to Shannon's formula for the amount of information required to describe the state of the gas, $I = -\sum_j p_j \log_2 p_j$, we see that the entropy of the gas S is equal to $k_B \ln 2\, I$. Equivalently, as noted by Zurek (1989), the entropy of the gas in state j can be identified as $k_B \ln 2$ times the algorithmic information content $K(b_i)$ of the bit string b_j describing the state j. Since, as noted above, algorithmic information is a measure of randomness, identifying entropy as proportional to algorithmic information is consistent with Maxwell and Boltzmann's notion that entropy is a measure of disorder or randomness (Cover & Thomas, 1991; Zurek, 1989).

In a quantum computer a logical operation or "op" occurs when a quantum bit flips. For a general quantum system, we will say that an op occurs when a quantum degree of freedom jumps from one state to a distinguishable state (Lloyd, 2000, 2002). Examples of ops include

- an atom jumps from its ground state of energy to an excited state, or vice versa,
- an electron or nuclear spin flips,
- a particle moves from one site to another, or from one momentum state to another,
- a photon is emitted or absorbed by an atom,
- a protein changes its configuration while folding,
- molecules in a chemical reaction exchange molecular sub-units,

and so on. A quantum bit flip is an op, and for a quantum computer to simulate an op in a physical system, a quantum bit must flip. The energetics of quantum bit flips and ops are known: for a quantum system to move from one state to a distinguishable state, energy must be applied. If the amount of energy employed in performing the op is E, then the minimum time it takes to perform the op is $\Delta t = \pi\hbar/2E$, where $\hbar = 1.0546 \times 10^{-34}$ joule-seconds is Planck's reduced constant. This relationship – known as the Margolus–Levitin theorem (1998) – has a simple intuitive interpretation: the faster one wishes to flip a bit or to move an electron from here to there, the more energy must be applied (Lloyd, 2000, 2002).

By applying the straightforward relationships between information and entropy, and between bit flips and ops (quantum systems moving from one state to another), one can measure the number of bits registered and number of ops performed by any chunk of matter in any volume of space-time (Lloyd, 2000, 2002). For example, the matter and energy within the part of the universe to which we have observational access has performed at most 10^{123} ops on 10^{92} bits since the big bang (Lloyd, 2002). That is, the correspondence between physical dynamics and computation is not merely a mathematical fact, it is a physically measurable phenomenon.

5.3 HOW AND WHY PHYSICAL SYSTEMS PRODUCE COMPLEXITY

We are now in a position to state and to prove some fundamental theorems on the generation of complexity.

Theorem 5.1 (Solomonoff–Kolmogorov–Chaitin) Consider a bit string b. The probability that a universal Turing machine U programmed with a random program produces b is no less than $2^{-K_U(b)}$ and is approximately equal to this number.

Proof (Solomonoff, 1964; Chaitin, 1987; Kolmogorov, 1965; Li & Vitanyi, 2008) If the program is random, then the probability that the first $K_U(b)$ bits of the Turing machine's program tape contain the correct shortest program that produces b is $2^{-K_U(b)}$. The algorithmic probability that a random program for U produces b as output is simply $p_U(b) = \sum_{q:U(q)=b} 2^{-|q|} \geq 2^{-K_U(b)}$, where $\{q : U(q) = b\}$ is the set of programs q that produce b as output and $|q|$ is the length of q in bits. In fact, it is possible to prove that the sum is dominated by the program with the shortest length and $p_U(b) \approx 2^{-K_U(b)}$.

Since, as noted above, the algorithmic informations defined with respect to different computationally universal systems U and U' are equal to within an additive constant, the algorithmic probabilities defined with respect to U and U' are equal to within a multiplicative constant. Accordingly, if we can show that one computationally universal system, programmed with a random program, produces complex structures with a non-zero probability, then all computationally universal systems programmed with random programs produce complex structures with non-zero probability.

Theorem 5.1 implies that if a complex mathematical structure b – by any desired definition of complexity including logical or thermodynamic depth – can be produced by a program of finite length, then it will be produced with finite probability by a randomly programmed universal Turing machine. Structures with low algorithmic information content will be produced with higher algorithmic probability than structures with high algorithmic information content. Since, as noted above, the observable part of the universe is computationally equivalent to a universal quantum computer, Theorem 5.1 implies that as long as our universe is supplied with a suitable source of initial random information as program, it must necessarily produce complex structures with non-zero probability.

In fact, with quantum computers, the random information in the initial program can be supplied by quantum fluctuations. As will be shown below, quantum fluctuations do indeed form the source of random information in our universe.

The computational equivalence between physical laws and quantum computers then allows us to state a physical version of Theorem 1.

Theorem 5.2 (Lloyd, 1993, 1997) Any subsystem of the universe that possesses the following physical properties necessarily generates complex structures at a rate proportional to their algorithmic probability.

(1) The laws of physics in that subsystem support universal quantum computation.
(2) The subsystem is initially in a state of low entropy compared with its maximum entropy. That is, the system has access to a large number of bits of memory space in a simple initial state (e.g., $00\ldots0$).
(3) The system has access to sufficient sources of energy to perform a large number of ops. Low entropy energy is called "free energy". So (2) and (3) can be summarized by demanding that the system start out in a state of high free energy.
(4) The initial state of the degrees of freedom of the subsystem that correspond to the program inputs for universal computation are either (a) in a random state, or (b) in a uniform quantum superposition of all possible program states. In case (a) the system is programmed with a random program. In case (b) it is programmed by random quantum fluctuations.

Proof (Lloyd, 1993, 1997) Conditions (1)–(4) imply that the subsystem is computationally equivalent to a universal quantum Turing machine that begins with a random program state. The straightforward generalization of Theorem 5.1 to quantum Turing machines then implies that this subsystem generates mathematical structures with a probability proportional to their algorithmic probability. In particular, since there exist complex structures with low algorithmic complexity, the system must generate complex structures with non-zero probability.

The reader may object that the computationally universal system is most likely to generate structures with very low algorithmic information content, such as 00...0. This is true, but it is important to note that one of the shortest possible programs, corresponding to a high algorithmic probability, generates all computable structures either sequentially or in parallel. That is, this program generates $U(0)$, $U(1)$, $U(00)$, $U(01)$, $U(10)$, $U(11)$, $U(000)$,...$U(q)$...for all possible programs q (some of which may not give an output). That is, a very short program can generate an ensemble of different results, some of which individually possess high algorithmic information content. The ensemble can have much lower algorithmic information content than a typical member, in the same way that the set of billion-bit numbers has low algorithmic information content – a few hundred bits at most – while a typical billion-bit number has algorithmic information content on the order of a billion. The description of the whole can be simpler than the description of an individual part.

Theorem 5.1 is a half-century old result from the theory of algorithmic information (Solomonoff, 1964; Chaitin, 1987; Kolomogorov, 1965; Li & Vitanyi, 2008). Theorem 5.2 uses the mathematical correspondences listed above to map Theorem 5.1 to the domain of physical systems (Lloyd, 1993, 1996, 1997, 2000, 2002). Theorem 5.2 implies that any computationally universal physical system with access to large amounts of free energy and to a source of randomness for its program must necessarily generate structures – including structures of arbitrarily high complexity – at a rate proportional to their algorithmic probability.

5.4 QUANTUM FLUCTUATIONS AND DECOHERENCE

Before analyzing how the universe generates complexity, I will make a few clarifying remarks about quantum fluctuations. As noted above, quantum mechanics allows physical systems – including the universe as a whole – to exist in quantum superpositions of states with widely varying properties. If the dynamics of the universe or of one of its subsystems depends on the way in which quantum superpositions interfere with each other over time, then that dynamics is said to be "coherent". When superpositions cease to interfere with each other, the resulting dynamics is called "decoherent". There are two complementary theories of how decoherence takes place. The first is called environmentally induced decoherence (Zurek, 1991): this form of decoherence occurs when interactions between a subsystem and its environment cause the quantum degrees of freedom of subsystem and environment to become correlated. In particular, if two components of a superposition in the quantum state of some subsystem become correlated via interaction with degrees of freedom of the environment – so that the environment obtains a bit of information about the system – then the dynamics of the system on its own can not exhibit interference between those two components. When the environment decoheres a system in this fashion, a random bit of information is created (just as if the environment were a measuring apparatus that interacts with the system to distinguish between the two components of the superposition). This random bit persists, and the system remains decoherent, as long as future interactions between system and environment do not undo the correlation. Clearly, environmentally induced decoherence can't be applied to the universe as a whole, as – by definition – the universe has no environment. It can however be applied to pieces of the universe, such as the part of the universe within an observer's past light cone.

The second, complementary theory of decoherence is that of consistent or decoherent histories (Griffiths, 2002; Omnés, 1994; Gell-Mann & Hartle, 1993). The method of decoherent histories does not rely on environmental interactions, and can be applied to the universe as a whole. This method looks at histories of observable quantities over time (including over the entire history of the universe, past and future) and provides mathematical criteria for when those histories fail to interfere with each other. Non-interfering histories are called decoherent histories. The method of decoherent histories is compatible with environmentally induced decoherence: degrees of freedom that are decohered by interaction with the environment exhibit decoherent histories. In contrast to environmentally induced decoherence, however, the decoherent histories method can be applied to infinite systems without environment, such as the universe.

The quantum fluctuations that give rise to structure and complexity in the universe are components of quantum superpositions that decohere. Decoherence injects bits of random information into the universe, which then become the seeds for future structure. Let's look at this process now in detail.

5.5 COMPLEXITY GENERATION IN THE UNIVERSE – PAST AND FUTURE

I now apply the results derived above to give a history of the generation of complexity in the universe, and to extrapolate that complexity generation to the future. First, I focus on the observationally accessible universe within the particle horizon, as the physics of this part of the universe is relatively well-established. Second, I investigate the history and future of complexity generation in the type I multiverse – the infinite space beyond the horizon governed by the same laws of physics as within the horizon, but specified by different initial conditions and quantum fluctuations.

Because inflation spreads out initially localized quantum fluctuations in density at an exponential rate, it produces a "scale-free" or power-law distribution of density fluctuations (Liddle & Lyth, 2000; Mukhanov, 2005; Weinberg, 2008; Birrell & Davies, 1982; Guth, 1981). The spreading out of the fluctuations means that parts of the wave function that were initially interacting no longer interact, and become "frozen in". Comparing with the theory of environmentally induced decoherence, we see that such large-scale quantum fluctuations are decoherent, and form the seeds for large-scale fluctuations in the cosmic microwave background observed in the current era. These scale-free fluctuations represent the injection via quantum decoherence of a large number of additional bits specifying the resulting matter and energy density.

As the wave function reaches the bottom of the local energy landscape, the energy in the vacuum thermalizes and creates a large number of particles at high temperature. This is the beginning of the hot big bang. Decoherence occurs because the different degrees of freedom in the universe interact strongly and spread energy rapidly throughout the system. The amount of information injected by decoherence at this stage dwarfs all amounts injected earlier. Observations of the cosmic microwave background show that thermalization and resulting decoherence is highly efficient: the temperature is almost constant throughout the sky, with deviations arising from the frozen-in remainder of the initially scale-free fluctuations arising from inflation. The number of bits injected into a chunk of matter/energy with volume V and temperature T is just the entropy of that chunk divided by $k_B \ln 2$. Indeed, this is the point at which most of the ca. 10^{92} bits in the matter in the observable universe were created.

At this point – the beginning of the hot big bang – one might think that everything should be over. After all, thermalization takes the matter to its maximum entropy state. Accordingly, the matter on its own has essentially no free energy available to store and flip bits in an extended computation. So it might appear that we can't apply Theorem 5.2. Even though the matter itself is in a state of essentially zero free energy, however, the matter taken together with the gravitational field is in a state of very high free energy. Due to a peculiar feature of gravitational systems, a completely homogeneous state of matter in a flat gravitational background – i.e., the beginning state of the hot big bang – is a state of minimum entropy and maximum free energy (Liddle & Lyth, 2000; Mukhanov, 2005; Davies, 1974) (see also discussions by Davies and Lineweaver in this volume). As noted by Newton (1687), the homogeneous state is unstable: tiny perturbations in density, such as the scale-free perturbations created by inflation, grow under the action of the gravitational force. The information injected into the universe via decoherence is transformed and processed by the laws of physics to amplify tiny structures to large scales – this process of amplification can be thought of as the initial computation performed by the computationally capable universe.

The computation that the universe performs at this era is a process in which entropy and algorithmic information are generated by quantum dynamics and decoherence. The gravitional instability of the flat universe processes and amplifies this information, "infecting" the gravitational field with randomly created seeds for large-scale structure. It is important to note that the precise scientific description of how quantum fluctuations in matter induce correlations in a quantized gravitational field must await a detailed theory of quantum gravity.

As the universe cools down, different species of particles – electrons, protons, helium nuclei, etc. – start to condense out of the still very hot primordial stew (Liddle & Lyth, 2000; Mukhanov, 2005; Weinberg, 2008). Until ca. 377 000 years after the big bang, matter and energy form a plasma of light nuclei, electrons, and photons. At this point, the universe becomes sufficiently cool for electrons to be bound to nuclei to form atoms (recombination). At the same time, the universe becomes transparent, as photons can now propagate long distances before scattering off those atoms. The photons of the cosmic microwave background date to this era.

All along, the gravitational force has been amplifying the primordial fluctuations from inflation, leading to the formation of stars, galaxies, clusters of galaxies, and superclusters: structures form on all scales. Early stars are made of light elements such as hydrogen and helium produced during the big bang: heavier nuclei are formed when these stars burn through their nuclear fuel and explode in supernovae. About four and a half billion years ago, our Solar System formed from primordial hydrogen and helium together with the remnants of such supernovae.

Up to the formation of second generation stars and planets, most of the structure formation has taken place either at large scales (gas giant planets, stars, galaxies, etc.) or tiny scales (thermal fluctuations in primordial gases). As heavier atoms clump together to form solid planets with water and atmospheres, conditions become right for complex chemical reactions to take place. The laws of chemistry are computationally universal – complex chemical reactions can perform any desired logical transformation. Large sources of free energy existed in the form of concentrations of atomic species created by supernovae, churned by convective and tidal forces within planets, heated by stars and cooled by the night sky. Planets, comets, and asteroids presented a wide variety of different environments – this variety arose from the varied dynamics of planetary formation acting on different initial conditions. The different initial conditions represented information arising by quantum fluctuations and injected into the universe via decoherence.

Clearly, at this point, Theorem 5.2 kicks in: the different planetary environments and initial conditions form the program whose information is processed by the computationally universal laws of physics and chemistry using the large amounts of available free energy – the raw material needed to perform ops on bits. By Theorem 5.2, complex structures necessarily arose. Somewhere, somehow, some of these complex structures formed the basis for primitive life – systems whose structure was encoded in a systematic form, and that could reproduce themselves with variation.

Once such self-reproducing proto-life forms, all bets are off. The encoded proto-genetic material becomes the program which – implemented by the laws of chemistry using readily available free energy – gives rise to the next generation of living systems with variations in their genetic material. As usual, variations in the information encoded by the genetic material can be traced to variations in environment and to mutations – all such variations can be traced backwards eventually to the quantum fluctuations that gave rise to them. (As noted above, quantum fluctuations, locked in by decoherence, are the only true source of undetermined variation in an otherwise deterministic world.) Indeed, once it showed up on Earth, life rapidly took over, evolving to take advantage of a wide variety of sources of free energy, and reproducing to the extent that it caused dramatic global changes in atmospheric composition and climate.

The rapid evolution of life on Earth has obscured its own origins. Fossil evidence shows that prokaryotic cells (simple cells lacking a nucleus) have existed for at least 3.5 billion years. Since the Earth developed a hard crust only 4.5 billion years ago, life apparently appeared on Earth shortly after conditions allowing it arose. Fossil evidence suggests that more complicated eukaryotic cells (cells with nuclei, many of which are capable of banding together to form multicellular organisms) arose at least 1.6–2.1 billion years ago. Once again, the more complicated forms of information processing rapidly took advantage of newly available forms of free energy to spread dramatically, leading up to the Cambrian explosion (ca. 530 million years ago), a reproductive information-processing fest that gave rise to major phyla of life.

A more complete catalogue of the sequence of information processing revolutions in the universe can be found in Lloyd (2006). All of these revolutions arise, at bottom, from the dynamic prescribed by Theorem 5.2. Information is injected into the universe via quantum fluctuations and decoherence; this information is itself random, but provides programs for the computationally universal laws of nature to process and transform this information in ways that construct complex structures. The existence of memory space and energy to perform ops is guaranteed by the presence of sources of free energy in the universe that arise, at bottom, from the interaction between matter and gravitational degrees of freedom.

5.6 GENERATION OF COMPLEXITY IN
THE MULTIVERSE

The universe is generally considered to consist of all that is. As noted above, the situation is somewhat more complicated in that – depending on the physical model used – there are various types of multiverse, each containing a large or potentially infinite number of universes (Tegmark, 2007). We have discussed the type I multiverse – the infinite volume beyond our horizon, in which each finite-volume set of initial conditions is repeated an infinite number of times. This first type of multiverse is nothing more or less than the standard inflationary cosmological model, amply supported by the combination of observation and theory. The type II multiverse consists of the connected set of "bubble" universes created by eternal inflation, where each bubble – itself infinite in spatial extent – can be governed by different laws of physics. But these are not the only multiverses in Tegmark's classification. In the quantum mechanical theory of the universe used here, the observed universe makes up just one component in the quantum mechanical superposition of all possible universes – the observed universe is the component picked out by the information injected into the universe by quantum fluctuations combined with decoherence. In quantum mechanics, the superposition of all components – or of all possible universes – has long been called the multiverse. In Tegmark's classification, this superposition of different quantum universes is called the type III multiverse. For adherents of the many worlds theory of quantum mechanics (Nielsen & Chuang, 2000; Peres, 2002), these other universes in the superposition are the other worlds in the many worlds. Because of decoherence, the other components in the superposition can have no effect on our component: they make up the part of the wave function of the multiverse that is inaccessible to our observations. Indeed, they make up the part of the wave function in which those observations turned out differently.

In fact, as noted by Aguirre *et al.* (2010), Nomura (2011), and Bousso and Susskind (2011), the quantum multiverse of multiple universes in superposition – which has long been controversial in the guise of the many worlds theory – becomes almost prosaic when combined with the type I and type II multiverse theories. Under rather mild assumptions, the fraction of universes that possess particular initial conditions (type I) and sets of physical laws (type II) is simply proportional to the square of the amplitude for such universes in the quantum multiverse (type III). This nifty trick allows a kind of "multiverse unification". In all types of multiverse, the combination of an infinite number of universes together with computational universality implies that somewhere, somehow, whatever can happen will happen. Indeed, all computable sequences of events of non-zero measure happen an infinite number of times.

5.7 GENERATION OF COMPLEXITY IN THE TYPE I MULTIVERSE

The type I multiverse consists of the infinite volume of space within and beyond our horizon. For convenience, imagine that this space is divided up into volumes of the same size as the volume within our horizon. Let's look at the generation of complexity in all such volumes, and how it differs from that within our horizon.

First of all, since the laws of physics are the same in all such volumes, and since large amounts of free energy are available, Theorem 5.2 kicks in and guarantees that a non-zero fraction of these volumes will generate structures of high complexity. How do these structures differ? First of all, while the initial spectrum of fluctuations obeys the same scale-free form everywhere, the specific configuration of fluctuations in energy density differs from volume to volume. In the great majority of volumes, the distribution of planets, stars, galaxies, and clusters of galaxies looks qualitatively similar to that in our volume, but the actual relative positions, masses, etc. of the components of these "universes" differs from ours in all but a tiny fraction.

In addition to such typical volumes, which look much like our universe, there is a small but finite fraction of universes with distributions of matter and energy that differ dramatically and qualitatively from ours. These atypical universes within the type I multiverse can exhibit a wide variety of exotic distributions.

As the history of the type I multiverse continues, more quantum accidents accumulate: the thermal fluctuations after re-thermalization translate into different microscopic fluctuations in the microwave background. Perhaps more striking is what happens when proto-life and life begin to develop. Mutations are quantum accidents, frozen in by decoherence. Reproduction under natural selection amplifies favorable mutations: a mutation that confers a 1% reproductive advantage will spread and dominate a population after more than one hundred generations. While we don't know the degree of dependency of the tree of life on exactly which mutations occurred when, the amplification effect of natural selection makes this process uniquely sensitive to quantum accidents via, for example, damage to DNA by cosmic rays. If life came into existence independently on some other planet within our universe, or beyond our horizon, that life might operate by substantially different biological mechanisms from life on Earth.

Now consider universes within the type I multiverse that progress identically with our universe up to the time of appearance of human beings on the scene (recall that there are an infinite number of such universes, representing a small but finite fraction of all the universes in the type I multiverse). Human history is full of dependence on various accidents. Every time some accident arises in part from the amplification of a tiny event, the accident can be traced back to a quantum fluctuation, frozen in by decoherence. The capricious nature of weather has affected many a human endeavor, for example. Because of the chaotic nature of the dynamics of the atmosphere – the butterfly effect – the intrinsic unpredictability of the weather arises at bottom from quantum fluctuations and decoherence. Molecular accidents during recombination during the process of sexual reproduction have a similarly quantum mechanical nature. The type I multiverse contains universes in which all such possible accidents play out their consequences. How would history have differed if Napoleon had been two meters tall, for example? Somewhere in the multiverse, people know.

5.8 GENERATION OF COMPLEXITY IN THE TYPE II MULTIVERSE

Eternal inflation gives a wider scope to the generation of complexity. The different bubbles within the type II multiverse explore not just all possible initial conditions, but a wide variety of different possible laws of physics. Just which such laws are explored depends on the details of the particular model of eternal inflation used. In type II universes, for example, living entities might sample the joys of life in two spatial dimensions, as in Abbott's Flatland (Abbott, 1992), or in four or more. They might live on space-time manifolds with two temporal dimensions, so one could maneuver around events in time – whatever that might mean! The sky is the limit.

5.9 GENERATION OF COMPLEXITY IN THE TYPE III MULTIVERSE

As noted above, in an infinite universe that explores all possible quantum fluctuations, the superposition of universes that makes up the type III multiverse can be indistinguishable – from the point of view of an observer living in the multiverse – from a type I or type II multiverse (Aguirre *et al.*, 2010; Nomura, 2011; Bousso & Susskind, 2011).

5.10 GENERATION OF COMPLEXITY IN THE TYPE IV MULTIVERSE

The final multiverse proposed by Tegmark (2007, 1998, 2008), is the type IV multiverse consisting of all computable physical theories and the universes governed by them. A classical version of this theory has been proposed by Schmidhuber (1997). This Platonic ideal of the set of universes as corresponding to the set of computable mathematical structures is certainly appealing, as it contains essentially every possible mathematical structure that can be specified using a finite amount of information (Turing, 1937; Papadimitriou & Lewis, 1982). This proposal is consistent with the observation that the known laws of our universe are computable. Its only drawback is that the correspondence between universes and computable mathematical structures is not quite clear. For example, is the set of integers together with addition and multiplication a "universe"?

5.11 RETURN TO OUR UNIVERSE

In fact, as I will now show, the Platonic type IV universe may well be contained in the future history of our own universe within the horizon. In its own humble way, our universe is potentially capable of generating all possible computable mathematical structures as summarized in the requirements of Theorem 5.2 (Lloyd, 1993, 1997, 2006). That is, the laws of the universe within the horizon are computationally universal; quantum fluctuations program our universe with all possible programs; and free energy is available to carry out those computations. The only open question that must be resolved before we can conclude that our universe will in fact eventually construct all computable structures is a physical one that is resolvable by a combination of theory and experiment. That question is whether free energy will continue to be available in the future. This question arises from a subtlety in the definition of a universal Turing machine: although a universal Turing machine is finite, it always has access to additional squares of blank tape, if needed. For a finite physical system – such as the volume of space within the particle horizon at time t since the big bang – to be computationally universal, it must have access to new bits of memory space and energy to perform ops for arbitrarily long times in the future. That is, if the universe is to compute forever, the different pieces of the universe must have ongoing access to free energy.

Whether or not free energy will continue to be available within the observable universe for all time is an open physical and cosmological question. If the observed cosmological term that is causing the accelerating expansion of the universe remains bounded away from zero (Riess *et al.*, 1998; Perlmutter *et al.*, 1999; Tegmark *et al.*, 2004), then the answer is apparently No. In this case, there will always be a de Sitter horizon at a finite distance, limiting the accessible volume of space. The maximum number of bits available for computation is limited by the maximum entropy within the horizon, equal to the horizon area divided by 4 ln 2 times the Planck length squared. Similarly, the amount of free energy that can be scavenged from distant pieces of the universe goes to zero, as the universe within the horizon has a minimum temperature which will eventually prevent waste heat from computation from being discarded. The universe in which the cosmological term remains bounded away from zero will eventually be a dismal and dull place in which nothing happens for very very long periods of time.

There exists a possible rescue from the purgatory of a non-zero cosmological term. Due to a highly unlikely statistical fluctuation, a low-entropy state (Davies, 1974) (e.g., a so-called "Boltzmann brain" (Dyson *et al.*, 2002; Albrecht & Sorbo, 2004; Linde, 2007)) can spontaneously materialize out of a high-entropy state. The probability of such a fluctuation occurring during the time it takes light to traverse the compass of the universe from horizon to horizon goes as $e^{-\Delta S}$, where ΔS is the magnitude of the entropy decrease. The entropy decrease required to assemble an actual brain out of thermal equilibrium is on the order of the number of elementary particles in the brain (ca. 10^{30}). So you would have to wait for light to cross the universe around $e^{10^{30}}$ times for your brain to reconstitute. Don't hold your breath.

By contrast, if the cosmological term goes to zero in the limit that $t \to \infty$, then the distance to the particle horizon also goes to ∞ in this limit, as does the number of bits and quantity of free energy available for computation. In this case the answer to the question is Yes: our universe is indeed capable of universal computation within the ever expanding particle horizon and can perform such computations for ever. All computable structures will eventually be created.

(Just how they might be created is described in detail in Lloyd (2006).) They will either be produced within our horizon or will enter our horizon from another part of the universe. Unlike the purely Platonic universes envisaged in Tegmark (2008) and Schmidhuber (1997), if the universe can compute for ever, then it will eventually produce and encompass all computable structures. We can only hope.

ACKNOWLEDGEMENTS

This work was supported by the Santa Fe Institute under a Miller Fellowship, NSF, DARPA, Lockheed Martin, Intel, ARO under a MURI program, NEC, and Jeffrey Epstein.

REFERENCES

Abbott, E. A. (1992). *Flatland: a Romance in Many Dimensions* (1884). New York: Dover.

Aguirre, A., Tegmark, M., & Layzer, D. (2010). Born in an infinite universe: a cosmological interpretation of quantum mechanics. arXiv:1008.1066.

Albrecht, A. & Sorbo, L. (2004). Can the universe afford inflation? *Phys. Rev. D.*, **70**, 063528.

Bennett, C. H. (1985). Dissipation, information, computational complexity and the definition of organization. In D. Pines (ed.), *Emerging Syntheses in Science*. Redwood City CA: Addison-Wesley, pp. 215–233.

Bennett, C. H. (1990). How to define complexity in physics, and why. In W. H. Zurek (ed.) *Complexity, Entropy and the Physics of Information*. Redwood City CA: Addison-Wesley, pp. 137–148.

Birrell, N. D. & Davies, P. C. W. (1982). *Quantum Fields in Curved Space*. Cambridge: Cambridge University Press.

Bousso, R. & Susskind, L. (2011). The multiverse interpretation of quantum mechanics. arXiv:1105.3796.

Chaitin, G. J. (1987). *Algorithmic Information Theory*. Cambridge: Cambridge University Press.

Cover, T. M. & Thomas, J. A. (1991). *Elements of Information Theory*. New York: Wiley.

Davies, P. C. W. (1974). *The Physics of Time Asymmetry*. Berkeley: University of California Press.

Dyson, L., Kleban, M., & Susskind, L. (2002). Disturbing implications of a cosmological constant. *J. High. Ener. Phys.*, **0210**, 011.

Gell-Mann, M. & Hartle, J. B. (1993). Classical equations for quantum systems. *Phys. Rev. D*, **47**, 3345–3382.

Griffiths, R. (2002). *Consistent Quantum Theory*. Cambridge: Cambridge University Press.

Guth, A. H. (1981). Inflationary universe: a possible solution to the horizon and flatness problems. *Phys. Rev. D*, **23**, 347–356.

Kolmogorov, A. N. (1965). Three approaches to the quantitative definition of information. *Prob. Inf. Trans.* **1**, 1–11.

Li, M. & Vitanyi, P. (2008). *An Introduction to Kolmogorov Complexity and Its Applications*, 2nd edn. New York: Springer-Verlag.

Liddle, A. R. & Lyth, D. H. (2000). *Cosmological Inflation and Large-Scale Structure*. Cambridge: Cambridge University Press.

Linde, A. D. (1986a). Eternally existing self-reproducing chaotic inflationary universe. *Phys. Lett. B*, **175**, 395–400.

Linde, A. D. (1986b). Eternal chaotic inflation. *Mod. Phys. Lett. A*, **1**, 81.

Linde, A. (2007). Sinks in the landscape, Boltzmann brains, and the cosmological constant problem. *J. Cos. Astr. Phys.*, **0701**, 022.

Lloyd, S. (1993). Quantum computers and uncomputability. *Phys. Rev. Lett.* **71**, 943–946.

Lloyd, S. (1995). Almost any quantum logic gate is universal. *Phys. Rev. Lett.*, **75**, 346–349.

Lloyd, S. (1996). Universal quantum simulators. *Science*, **273**, 1073–1078.

Lloyd, S. (1997). Universe as quantum computer. *Complexity*, **3/1**, 32–35.

Lloyd, S. (2000). Ultimate physical limits to computation. *Nature*, **406**, 1047–1054.

Lloyd, S. (2002). Computational capacity of the universe. *Phys. Rev. Lett.*, **88**, 237901.

Lloyd, S. (2006). *Programming the Universe*. New York: Knopf.

Lloyd, S. & Pagels, H. (1988). Complexity as thermodynamic depth. *Ann. Phys.*, **188**, 186–213.

Margolus, N. (1984). Physics-like models of computation. *Physica D*, **10**, 81–95.

Margolus, N. & Levitin, L. B. (1998). The maximum speed of dynamical evolution. *Physica D*, **120**, 188–195.

Messiah, A. (1999). *Quantum Mechanics*. New York: Dover.

Mukhanov, V. (2005). *Physical Foundations of Cosmology*. Cambridge: Cambridge University Press.

Newton, I. (1687). Philosophiae Naturalis Principia Mathematica. London: Royal Society.

Nielsen, M. A., Chuang, I. L. (2000). *Quantum Computation and Quantum Information*, Cambridge: Cambridge University Press.

Nomura, Y. (2011). Physical theories, eternal inflation, and the quantum universe. arXiv:1104.2324.

Omnés, R. (1994). *The Interpretation of Quantum Mechanics*. Princeton: Princeton University Press.

Papadimitriou, C. H. & Lewis, H. (1982). *Elements of the Theory of Computation*. Englewood Cliffs: Prentice-Hall.

Peres, A. (2002). *Quantum Theory: Concepts and Methods*. New York: Kluwer.

Perlmutter, S., Alderling, G., Goldhaber, G. *et al.* (1999). Measurements of omega and lambda from 42 high-redshift supernovae. *Astro. Journal*, **517**, 565–586.

Riess, A., Filippeuko, A. V., Challis, P. *et al.* (1998). Observational evidence from supernovae for an accelerating Universe and a cosmological constant. *Astro. Journal*, **116**, 1009–1038.

Schmidhuber, J. (1997). A computer scientist's view of life, the universe, and everything. In C. Freksa (ed.). *Foundations of Computer Science: Potential – Theory – Cognition*, Lecture Notes in Computer Science. New York: Springer, pp. 201–208.

Solomonoff, R. J. (1964). A formal theory of inductive inference, Part I. *Inf. Control*, **7**, 1–22. A formal theory of inductive inference, Part II. *Inf. Control*, 7, 224–254.

Starobinsky, A. A. (1982). Dynamics of phase transition in the new inflationary universe scenario and generation of perturbations. *Phys. Lett. B*, **117**, 175–178.

Susskind, L. (2007). The anthropic landscape of string theory. In B. Carr (ed.), *Universe or Multiverse*. Cambridge: Cambridge University Press.

Tegmark, M. (1998). Is 'the theory of everything' merely the ultimate ensemble theory? *Ann. Phys.*, **270**, 1–51.

Tegmark, M. (2007). The multiverse hierarchy. In B. Carr (ed.), *Universe or Multiverse*. Cambridge: Cambridge University Press.

Tegmark, M. (2008). The mathematical universe. *Found. Phys.*, **38**, 101–150.

Tegmark, M., Strauss, M. A., Blanton, M. R. *et al.* (2004). Cosmological parameters from SDSS and WMAP. *Phys. Rev. D*, **69**, 103501.

Turing, A. M. (1937). On computable numbers, with an application to the *Entscheidungsproblem*, *Proc. Lond. Math. Soc.*, **2**, 42, 230–265. Turing, A. M. (1937). On computable numbers, with an application to the *Entscheidungsproblem:* a correction. *Proc. Lond. Math. Soc.*, **2**, 43, 544–546.

Vilenkin, A. (1994). Topological inflation. *Phys. Rev. Lett.*, **72**, 3137–3140.

Weinberg, S. (2008). *Cosmology*. Oxford: Oxford University Press.

Wolfram, S. (1986). *Theory and Applications of Cellular Automata*, Advanced series on complex systems 1. Singapore: World Scientific.

Zurek, W. H. (1989). Algorithmic randomness and physical entropy. *Phys. Rev. A*, **40**, 4731–4751.

Zurek, W. H. (1991). Decoherence and the transition from quantum to classical. *Phys. Today*, **44**, 36–44.

6 Emergent spatiotemporal complexity in field theory

Marcelo Gleiser

The origin of spatiotemporal order in physical and biological systems
is a key scientific question of our time. How does microscopic matter self-organize to create living and non-living macroscopic structures? Do systems capable of generating spatiotemporal complexity
obey certain universal principles? We propose that progress along
these questions may be made by searching for fundamental properties of non-linear field models which are common to several areas
of physics, from elementary particle physics to condensed matter and
biological physics. In particular, we've begun exploring what models
that support localized coherent (soliton-like) solutions – both time-
dependent and time-independent – can tell us about the emergence of
spatiotemporal order. Of interest to us is the non-equilibrium dynamics of such systems and how it differs when they are allowed to interact with external environments. It is argued that the emergence of
spatiotemporal order delays energy equipartition and that growing
complexity correlates with growing departure from equipartition. We
further argue that the emergence of complexity is related to the existence of attractors in field configuration space and propose a new
entropic measure to quantify the degree of ordering of localized energy
configurations.

6.1 SOLITONS AND SELF-ORGANIZATION

A key question across the natural sciences is how simple material
entities self-organize to create coherent structures capable of complex
behavior. As an example, phenomena as diverse as water waves and

Complexity and the Arrow of Time, ed. Charles H. Lineweaver, Paul C. W. Davies
and Michael Ruse. Published by Cambridge University Press.
© Cambridge University Press 2013.

symmetry-breaking during phase transitions can give rise to solitons, topologically or non-topologically stable spatially-localized structures ("energy lumps") that keep their profiles as they move across space. They beautifully illustrate cooperative behavior in Nature, that is, how interacting discrete entities work in tandem to generate complex structures that minimize energy and other physical quantities (Infeld & Rowlands, 2000; Walgraef, 1997; Cross & Hohenberg, 1993; Rajamaran, 1987; Lee & Pang, 1992). For these to exist, a dynamic compromise must be achieved between what could be called gathering and dispersive tendencies: while attractive interactions tend to collect particles in small volumes, gradient energies want to spread them out. The study of such structures started in earnest in August 1834, when the Scottish engineer John Scott Russell was conducting experiments to improve the design of canals for boats. He observed, to his amazement, that sometimes, when the boat stopped suddenly, a "mass of water" kept travelling ahead of the boat for miles without losing its spatial shape. He called this "singular and beautiful phenomenon" a "Wave of Translation" (Scott, 2007). Today we call it a solitary wave or soliton.

In elementary particle physics, solitons owe their stability to a conserved charge, which can be either topological (that is, related to the non-trivial structure of the vacuum, i.e., the set of minima of the potential energy describing the system) or non-topological (that is, related to the conservation of certain charge-like quantities such as particle number). In both cases, solitons are time-independent solutions of the non-linear partial differential equations describing the models, which are robust against perturbations. They can be thought of as being composed of a superposition of many different momentum modes. Semi-classically, they can be thought of as collections of particles trapped in a bag-like configuration. It turns out that the same, or qualitatively similar, PDEs appear in many different areas of physics, albeit often in different spatial dimensionalities and in non-relativistic and/or high-viscosity approximations. This is due to the fact that many of the equations describing the interactions of fundamental matter fields are essentially non-linear wave equations with amplitude-dependent non-linearities determined by the particular interactions of the model. For example, the famous sine-Gordon equation (when $\sin x$ is truncated to second order, as in $\sin x = x - x^3/3!$) is related to the 1d Klein–Gordon equation with a double-well potential widely used in describing symmetry breaking in particle physics and condensed matter theories (Rajamaran, 1987). Since the 1970s, a large amount of literature exploring the properties of such solitons has been produced. The focus is invariably the same: a given model is described by a given PDE or coupled PDEs. Static solutions to these equations describe localized, coherent spatial structures such as kinks, vortices, or monopoles. In some cases, these structures maintain their profiles after scattering (what are known as "real" solitons), while in others they may suffer perturbations but still approximately maintain their spatial coherence or combine into more complex hybrid objects.

One of the key points I'd like to make is that focusing only on static (time-independent) solutions, as has mostly been the case for the past four decades, adds an unnecessary and very stringent constraint on the classes of models that can exhibit interesting spatiotemporal coherent behavior. Once this constraint is relaxed and we look for *time-dependent* but still spatially-localized structures, a whole new world opens up, with a plethora of possibilities. As I have shown over the past 15 years with various collaborators, oscillating soliton-like structures appear in a wide class of fundamental models describing symmetry breaking in elementary particle physics, cosmology, and in condensed matter systems. I called them *oscillons* in 1994 (Gleiser, 1994; Copeland, Gleiser, & Muller, 1995). Since then, their remarkable properties have attracted much attention in high energy physics and cosmology (Gleiser & Sornborger, 2000; Honda & Choptuik, 2002; Adib, Gleiser, & Almeida, 2002; Graham & Stamatopoulos, 2006; Farhi *et al.*, 2008; Fodor *et al.*, 2006; Gleiser & Howell, 2005; Gleiser & Sicilia, 2008, 2009; Hertzberg, 2010; Amin & Shirokoff, 2010; Amin *et al.*, 2012). Interestingly, also in 1994, spatiotemporal patterns emerging in vibrating grains have been discovered and also given the name oscillons (Melo *et al.*, 1994; Umbanhowar *et al.*, 1996; Tsimring & Aranson, 1997; Jeong & Moon, 1999). Similar localized oscillating structures have been found in the plasma of stellar interiors (Umurhan *et al.*, 1998), in extensions of the Swift–Hohenberg model (Crawford & Riecke, 1999), in deep water vibrations (Shats, Xia, & Punzmann, 2012), and in the non-linear Schrödinger equation (Stenflo & Yu, 2007), to list but a few examples. They all seem to have very similar qualitative properties to fundamental oscillons, although a more detailed analysis should be done to establish their correspondence.

6.2 EMERGENT SPATIOTEMPORAL COMPLEXITY IN FIELD THEORY

I will use the name *oscillons* to characterize any long-living, oscillating coherent field configuration found in non-linear field models in arbitrary spatial dimensions. These can involve a single real scalar field (Bogolubsky & Makhankov, 1976; Gleiser, 1994; Copeland, Gleiser, & Muller, 1995; Gleiser & Sornborger, 2000; Honda & Choptuik, 2002; Adib, Gleiser, & Almeida, 2002; Graham & Stamatopoulos, 2006; Farhi *et al.*, 2008; Fodor *et al.*, 2006; Gleiser & Howell, 2005; Gleiser & Sicilia, 2008, 2009; Hertzberg, 2010; Amin & Shirokoff, 2010, Amin *et al.*, 2012) or several interacting fields (Gleiser & Thorarinson, 2007, 2009). In discrete lattice models, qualitatively similar configurations are known by the name of *discrete breathers* (Flach & Willis, 1998). The name breathers also describes bound states of 1d kink–antikink pairs, which are localized in space and oscillate in time, somewhat like a drumhead that wouldn't stop vibrating: the non-linear interactions restrict the radiation of energy to spatial infinity (Gleiser & Sicilia, 2008).

In a sense, oscillons are the answer to Derrick's theorem (Rajaraman, 1987), which forbids the existence of time-independent solitons in more than one spatial dimension for models featuring only a single real scalar field. This disappointing result can be evaded in two ways: either by searching for static solutions of models combining more than one scalar field, as is the case of vortices (combining scalar and gauge fields) and non-topological solitons (combining real and complex scalar fields or just complex scalars, as in Q-balls) (Lee and Pang, 1992), or by having long-lived *time-dependent* coherent solutions, the oscillons. So, I propose that oscillons open a whole new chapter in the study of soliton-like configurations in field theories, which need not be time-independent to be physically interesting. As long as their lifetimes are longer than the typical timescales of the systems where they appear, they may have very important dynamical consequences.

Oscillons have been found numerically in two ways. Initially, they were found as spherically-symmetric solutions of the Klein–Gordon equation with symmetric and asymmetric double-well potentials (Gleiser, 1994; Copeland, Gleiser, & Muller, 1995). To find them, one simply starts with a localized large-amplitude perturbation, for example, an initial Gaussian profile, $\varphi(r, 0) = A_0 \exp[-r^2/R_0^2] - \varphi_v$, where A_0 and R_0 are the initial amplitude and radius, respectively. φ_v stands for the vacuum, that is, the minimum of the potential energy. Thus, the initial perturbation can be interpreted as a non-perturbative vacuum fluctuation. We've shown that such initial profiles would evolve into an *attractor* configuration, the oscillon, as long as the core amplitude $(A_0 - \varphi_v)$ probed into the non-linear part of the potential (that is, the region where $V'' < 0$) *and* the initial radius was above a critical value (Copeland, Gleiser, & Muller, 1995). These results were then generalized to more complex models with several interacting fields. The characteristic property of oscillons is an approximately energy-conserving non-linear oscillation of the core amplitude of the field about the vacuum φ_v.

More interestingly, oscillons were also found with generic, noisy initial conditions in different models: they were shown to emerge spontaneously as a system attempts to regain thermal equilibrium after a sudden change in the nature of its interactions (Gleiser, 2004; Gleiser & Howell, 2003; Gleiser & Thorarinson, 2007, 2009).

(Specifically, from a change in potential from a single to a double-well, which tosses the initial state out of equilibrium.) This is an extremely significant result, showing that physical systems that support non-linear coherent structures will naturally produce them as they evolve toward their final equilibrium and hence maximum entropy state. This is true whether the system is closed (unable to exchange energy with the outside and hence energy-conserving) or open, although a more detailed comparison is lacking. As long as there is enough initial energy in the system, these attractor coherent structures emerge from arbitrary initial conditions. More remarkably, we have shown that they may emerge in near-perfect synchrony across the spatial volume: not only is each oscillon in itself an example of coherent spatiotemporal behavior, but there is also self-organized behavior in time at large spatial scales. (This is due to our initial set up, as the synchronous behavior is induced by the oscillations of the zero-mode of the field. More on this later.) Figure 6.1 shows periodic snapshots of a two-dimensional simulation where oscillons of the field $\varphi(x, y, t)$ emerge together across the lattice, while Fig. 6.2 shows the synchronous emergence.

In practice, such structures can be modeled in three simple steps: first, prepare the system in an initial thermal state by solving the stochastic Langevin–Klein–Gordon equation of a scalar field in contact with a heat bath (the external environment) at temperature T:

$$\ddot{\varphi} - \nabla^2 \varphi + \gamma \dot{\varphi} = -V' + \xi(t, \vec{x}), \tag{6.1}$$

where γ is a viscosity coefficient that couples the system to the thermal bath (the "environment") and $\xi(t, \vec{x})$ is a stochastic force with zero mean and two-point correlation function given by the fluctuation–dissipation theorem as $\langle \xi(t, \vec{x}) \xi(t', \vec{x}') \rangle = 2\gamma T \delta(x - x') \delta(t - t')$. This Langevin–Klein–Gordon equation is a generalization of Brownian motion to a spatially-dependent field. Its role here is simply to drive the field into a thermal equilibrium state with temperature T. V' is the derivative of the potential with respect to the field. In order to

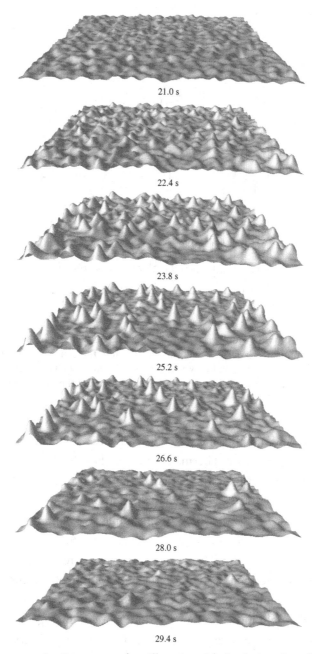

21.0 s

22.4 s

23.8 s

25.2 s

26.6 s

28.0 s

29.4 s

FIGURE 6.1 Emergence of oscillons in a 2d simulation. Snapshots are taken periodically. The vertical axis shows the magnitude of the scalar field $\varphi(x, y, t)$, while the horizontal axes are for the spatial coordinates x and y. Notice that while each oscillon is a coherent oscillating spatiotemporal structure, oscillons emerge in near-perfect phase across the lattice, showing large-scale self-organization (Gleiser & Howell, 2003). This can be seen in Fig. 6.2, where the number of oscillons nucleated is counted as a function of time.

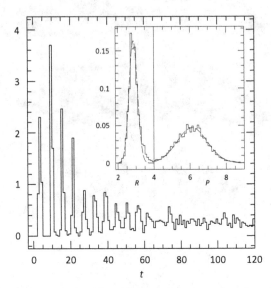

FIGURE 6.2 Number of oscillons nucleated as a function of time in a 2d simulation (Gleiser & Howell, 2003). Note the synchronous emergence for early times, and how it is lost for $t > 60$ or so. The inset shows the average radius (R) of the localized structures and their oscillation periods (P).

bring the field to a thermal state, it's convenient to choose $V = \varphi^2/2$, a simple linear theory. Thermalization can be verified by checking if the equipartition theorem is satisfied, that is, if $\varphi^2/2 = kT$. Once it is (to a desired accuracy), we switch the potential to a non-linear double-well, $V = \varphi^2/2 - \alpha\varphi^3/3 + \varphi^4/4$, for example. We can either keep the noise on (for an open system) or switch it off (for a closed, conservative system). Keeping it on means that we are investigating the dynamics of an open system, while switching it off means that we are investigating the dynamics of a conservative system. The quench throws the field into a non-equilibrium state, inducing energy transfer between the various non-linear modes. What we observe is what is depicted in Figs. 6.1 and 6.2, from a simulation for a 2d system. Localized spatial structures emerge spontaneously, driven mainly by large-amplitude oscillations of the longest wavelength mode (the "zero mode") of the system. While short wavelength modes remain in thermal equilibrium due to their fast relaxation rates, a band of long wavelength modes is excited via a mechanism known as "parametric resonance", which can be seen even in a simple harmonic oscillator with a periodically-driven frequency (Landau & Lifshitz, 1976; Gleiser & Howell, 2003). (A child being periodically pushed in a swing mimics this behavior quite nicely.)

In short, we have a physical system that displays complex spatiotemporal behavior as it is tossed out of equilibrium. The entities that emerge spontaneously are coherent time-dependent states of the scalar field, which are soliton-like even if time-dependent. In a sense, this behavior mimics some of the thermodynamic behavior you see in living systems, which are also self-organizing non-equilibrium dissipative structures, although living beings are of necessity open thermodynamic systems. We will explore this analogy further below. But first, we present a new measure of complexity in physical systems.

6.3 QUANTIFYING THE EMERGENCE OF COMPLEXITY: ORGANIZED STRUCTURES AS BOTTLENECKS TO EQUIPARTITION

Considering that the system is set up in thermal equilibrium initially, or at least with a certain mode distribution (it need not be thermal at all), the sudden change in the interactions sets it "out of equilibrium", or at least out of energy equipartition. There are many ways of achieving this. Here it was achieved by changing the field's potential. Since the physical system is now in an unstable state, it will evolve to reach a new equilibrium state. (We will consider only closed systems here.) It is during the early stages of this evolution to a new equilibrium state that self-organized oscillon-like patterns emerge (see Fig. 6.1). In fact, given that oscillons emerge nearly in phase (Fig. 6.2), they work as bottlenecks to equipartition, that is, they maintain the system out of equipartition longer than expected. In words, self-organizing structures lock within them long-wavelength modes, impeding them from interacting with shorter wavelength modes and thus from reaching equilibrium with the rest of the system. We can argue that the emergence of self-organizing spatiotemporal structures works like an entropy sink, at least in their immediate surroundings: although the total thermodynamic entropy always grows in accordance with the second law of thermodynamics, the presence of ordered structures slows down this progression, delaying equipartition.[1]

In the past, we proposed a possible measure for the emergence of ordered structures, $S(t) = -\int d^3k\, p(k, t)\ln[p(k, t)]$, where $p(k, t) = K(k, t)/\int d^3kK(k, t)$ is the fractional kinetic energy per mode and $K(k, t)$ is the kinetic energy of the k-th mode. Note that in the lattice (discrete) version of this expression, the maximum is given by $S_{\max} = \ln(N)$, where N is the number of degrees of freedom (Gleiser, 2004; Gleiser & Howell, 2003).

More recently, Gleiser and Stamatopoulos have developed a more efficient measure of this field entropy, which, for situations where the field is out of equilibrium, we call relative configurational entropy (Gleiser & Stamatopoulos, 2012a & 2012b). Essentially, it is inspired by the Kullback–Leibler divergence from information theory, adapted to field configuration landscapes. Further details are provided in Gleiser and Stamatopoulos (2012a, 2012b). We next describe how to compute the relative configurational entropy and illustrate it with an example. As will be clear, the configurational entropy offers a joint measure of the system's departure from equilibrium *and* of the emergence of spatially-coherent structures.

Given a field $\varphi(x, t)$ defined in a volume V, compute its Fourier transform at time t: $F(k, t)$. From it, define the modal fraction at time

---

[1] It is an interesting question whether ordered structures work to increase the total entropy in open thermodynamic systems. Is the total entropy of open systems with ordered structures larger than those without? For example, do hurricanes and living forms increase the total entropy of the Earth–Sun system? The answer is probably yes, since both ordered dissipative structures degrade solar energy (Prigogine, 1978; Prigogine, 1980; Lineweaver & Egan, 2008). Simulations of open systems with oscillons may allow one to study this issue in quantitative detail.

t, $f(k, t) = \frac{|F(k,t)|^2}{\int |F(k,t)|^2 d^d k}$, where d is the number of spatial dimensions. We are thus restricting either the field $\varphi(x, t)$ or, when appropriate, its energy density, to belong to the set of bound square-integrable functions. This makes sense since we are mainly interested in spatially-localized systems. We define the relative configurational entropy (RCE) as

$$S_f(t) = \int f(k, t) \ln \left[\frac{f(k, t)}{h(k)} \right] d^d k, \qquad (6.2)$$

where $h(k) \sim T/(k^2 + m^2)$ is the modal fraction for a thermal spectrum, with m the relevant mass scale. With this definition, the relative configurational entropy is a robust measure of the departure from thermal equilibrium in a physical system: if a system is thermalized, $f(k, t) = h(k)$, and the RCE vanishes.

In Fig. 6.3 we show the evolution of the RCE, $S_f(k, t) = f(k, t)\ln[f(k, t)/h(k)]$, computed for a 3d scalar field model right after the quench from a single to a double-well potential, as described above (Gleiser & Stamatopoulos, 2012b). Very rapidly, peaks develop in $S_f(k, t)$, illustrating the departure from equilibrium. Note that only low k modes get out of equilibrium, due to their longer equilibration timescale. These are the modes corresponding to the spatially-ordered structures in Fig. 6.1, the emerging oscillons. So, as the system is launched out of thermal equilibrium by the quench, it spontaneously generates spatiotemporal organization. This is a very clear example of complexity spontaneously emerging from self-interactions coupled to non-equilibrium conditions.

Our studies so far indicate that, at least in field theoretical models describing a wide number of physical phenomena, the rise of complexity is related to a strong departure from equipartition. So, although spatiotemporal complexity needs non-linearity as a prerequisite (see, e.g., Infeld & Rowlands, 2000), its spontaneous emergence needs environmental input. This input may be external (as in *really* environmental, such as a change in the temperature, in the value of an external field or a strong pressure perturbation), or internal (e.g., due

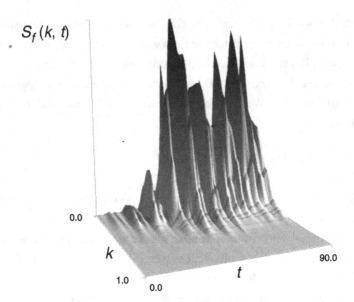

FIGURE 6.3 Time evolution of the RCE density for a real scalar field quenched from a single to a double-well potential in a 3d simulation. The peaks correspond to maximal oscillon formation, correlating with the synchronous emergence of ordered structures shown in Figs. 6.1 and 6.2. Note that only low k modes go out of equilibrium after the quench.

to a change in the system's self-interactions with time or with energy scale, as achieved here with a switch from a single to a double-well potential).

In order to compare the emergence of structure with the newly-introduced entropic measure, Gleiser and Stamatopoulos computed the (ensemble-averaged) number density of emerging oscillons as a function of time for different initial temperatures for the same 3d simulation (Gleiser & Stamatopoulos, 2012b). For small temperatures, no oscillons form, as should be expected: the field has small-amplitude oscillations about the minimum of the potential and won't probe its non-linear portion, where $V'' < 0$. As the temperature increases, a growing number of oscillons form after the quench. We found that they have a striking linear correlation with the RCE, as shown in

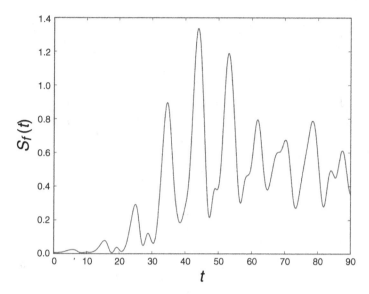

FIGURE 6.4 Entropic measure of spatiotemporal ordering due to oscillon emergence in 3d (Gleiser & Stamatopoulos, 2012b). The initial state $(t = 0)$ is thermal and thus of maximum entropy. We plot the momentum-integrated RCE as it evolves in time. The peaking occurs during maximal oscillon activity, as in Fig. 6.3. As the system evolves toward equipartition, the function approaches its asymptotic equilibrium value (not shown).

Fig. 6.5, where the two quantities are directly compared. *We thus propose that the relative configurational entropy can be used as a direct measure of complexity in the system, if by complexity we mean the emergence of coherent spatiotemporal order.*

6.4 INTO THE SPECULATIVE REALM: LIFE AS A SELF-SUSTAINING OSCILLON-LIKE STRUCTURE

A key question that remains unanswered is whether there are particular environmental interactions that help sustain spatiotemporal complexity in open dissipative systems against its natural tendency towards disappearance as the system evolves. This is what happens,

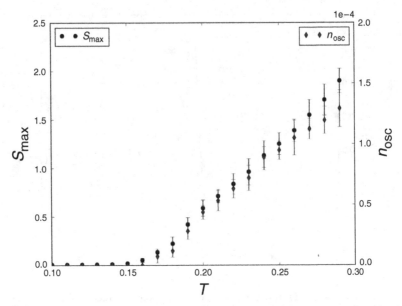

FIGURE 6.5 Ensemble-averaged values of the maxima of the RCE (dots) and number density of nucleated oscillons (diamonds) as a function of the initial temperature T. For $T < 0.15$ no oscillons form and the RCE is zero. For $T > 0.28$, the results start to diverge due to issues in the computation of the effective potential. For each temperature, results are averaged over 20 runs. Error bars are computed from this ensemble average (Gleiser & Stamatopoulos, 2012b).

for example, with hurricanes, where the fuel comes from the evaporation of warm tropical waters. In a sense, this is also what living systems do: they maintain their spatiotemporal order against its natural decay tendencies for much longer than expected due to their interactions with the environment. Can we study the essential physical mechanisms of living systems with non-linear field models of open dissipative systems (Prigogine, 1980; Lineweaver & Egan, 2008)?

Since the goal of this book is to try and break new ground on the study of complexity, I'd like to end my remarks with a somewhat daring suggestion: that some of the key physical mechanisms of living systems may be modeled with oscillon-like structures. We have seen that oscillons are long-lived coherent spatiotemporal structures that can interact with an external environment. They have nearly constant energy, although some of it leaks as they evolve to their demise. The study of how they interact with an external environment has barely begun: so far, all we have done is to study their evolution as open or closed thermodynamic systems, where in the "open" case we coupled them to an external heat bath model via a Langevin dynamics with Markovian noise (Gleiser, 2004; Gleiser & Howell, 2003). Most of the results in this chapter pertain to closed systems. In this case, we have seen that after some time the oscillons disappear as the system approaches thermal equilibrium. Nothing escapes the second law! For open systems, a violent environment will also destroy the field coherence. (Say, a heat bath with high temperature and viscosity, Gleiser & Haas, 1996.) However, beneficial couplings can also be contemplated, where the environment helps support the oscillon population. It is possible that particular modes of interaction will lead to an enhancement of lifetime and even to a complex dynamics of proliferation: oscillons can and do interact with each other (Hindmarsh & Salmi, 2006, 2008) and it is an open question whether their interactions may lead to spawning. It is well-known in the theory of phase transitions that nucleation often happens due to the presence of some impurity or perturbation in the system. Once nucleated, crystals and bubbles can grow, and sometimes they fission and coalesce, depending on the details of their mutual interactions and of external driving or inhibiting forces (Langer, 1992; Gunton, San Miguel, & Sahni, 1983; Gunton, 1999). At this point, we wouldn't have much of a genetic code analog, but could perhaps model "metabolism-first" systems (Dyson, 1984; Kauffman, 1993) in some detail: bubble-like resilient structures (protocells) are protected sites whereby more complex chemical reactions can occur. Some may accidentally contain pre-genetic material. Environmental fluctuations (such as turbulence) may induce a fraction of protocells to randomly split and combine with one another. Energy exchange with the environment may also help or hinder a protocell's ability to survive. Possibly, certain autocatalytic chemical reactions or other means may favor division and improve the protocell's robustness. Such entities would be favored over more unstable ones, eventually leading to a population of naturally-selected protocells as sites for prebiotic chemistry.

We have recently shown that certain non-catalytic prebiotic reaction–diffusion chemical networks may spontaneously form chiral protodomains immersed in an achiral substrate, thus generating spatiotemporal self-organizing structures (Gleiser & Walker, 2009b). Although the details of this work need to be explicitly connected with the emergence of spatiotemporal order discussed above, they both spring from the same source: non-linear PDEs describing non-equilibrium situations. We believe this effort may be an important new step connecting physics and biology through chemistry, with the emergence of complexity as the link across the disciplines.

6.5 FINAL REMARKS: LOOKING AHEAD

The advent of powerful desktop computers has permanently changed the scientific landscape. Problems that were once intractable are now routine. Non-linear systems are presently studied in great detail. In particular, parallel computing is extremely helpful in the implementation of the large-scale lattice simulations needed for investigating the non-equilibrium dynamics of classical (and quantum) fields. I have argued here that progress toward a more fundamental understanding of complexity will benefit from a detailed study of the emergence of spatiotemporal coherent structures in physical models inspired by elementary particles and condensed matter systems (Gleiser & Stamatopoulos, 2012a, 2012b). We have observed how such structures emerge spontaneously as the system is prepared in an unstable initial state. The irreversibility of time is of course related to this initial state: dynamical instability in systems with many degrees of freedom necessarily induces an arrow of time. That this dynamics also displays such rich spatiotemporal patterns as the system evolves to a final equilibrium state is due to the non-trivial action of non-linearities. We have further argued that the emergence of complexity is related to the existence of attractors in field configuration space and introduced a new entropic measure to quantify this emergence. As long as energy is supplied periodically, such structures remain active. Apart from providing insight into the dynamics of complexity, they may well serve as models for some of the basic thermodynamic properties of living systems.

REFERENCES

Adib, A., Gleiser, M., & Almeida, C. (2002). Long-lived oscillons from asymmetric bubbles. *Physical Review D*, **66**, 085011.

Amin, M. A. & Shirokoff, D. (2010). Flat-top oscillons in an expanding Universe. *Physical Review D*, **81**, 011602.

Amin, M. A., Easther, R., Finkel, H., Flauger, R. & Hertzberg, M. P. (2012). Oscillons after inflation. *Physical Review Letters*, **108**, 241302.

Bogolubsky, I. L. & Makhankov, V. G. (1976). *J. Exp. Theor. Phys. Lett.*, **24**, 12 [*Pis'ma Zh. Eksp. Teor. Fiz.*, **24**, 15].

Copeland, E. J., Gleiser, M., & Muller, H.-R. (1995). Oscillons: resonant configurations during bubble collapse. *Physical Review D*, **52**, 1920.

Crawford, C. & Riecke, H. (1999). Oscillon-type structures and their interaction in a Swift–Hohenberg Model. *Physica D*, **129**, 83.

Cross, M. C. & Hohenberg, P. C. (1993). Pattern formation outside of equilibrium. *Reviews of Modern Physics*, **65**, 851.

Dyson, F. (1984). *Origins of Life*. Cambridge, UK: Cambridge University Press.

Farhi, E., Graham, N., Guth, A. H. *et al.* (2008). Emergence of oscillons in an expand background. *Physical Review D*, **77**, 085019.

Flach, S. & Willis, C. R. (1998). Discrete breathers. *Physics Reports*, **295**, 181.

Fodor, G., Forgács, P., Grandclément, P. & Rácz, I. (2006). Oscillons and quasi-breathers in the phi^4 Klein–Gordon model. *Physical Review D*, **74**, 124003.

Gleiser, M. (1994). Pseudo-stable bubbles. *Physical Review D*, **49**, 2978.

Gleiser, M. (2004). The problem of the 3 origins: cosmos, life, and mind. In J. Barrow, P. C. W. Davies & C. Harper, Jr. (eds.), *Science and Ultimate Reality: a Celebration of John A. Wheeler's Vision*. Cambridge, UK: Cambridge University Press.

Gleiser, M. & Haas, R. (1996). Oscillons in a hot heat bath. *Physical Review D*, **54**, 1626.

Gleiser, M. & Howell, R. (2003). Resonant emergence of local and global spatiotemporal order in a nonlinear field mode. *Physical Review E*, **68**, 065203(RC).

Gleiser, M. & Sicilia, D. (2008). An analytic characterization of oscillons: their energy, radius, frequency, and lifetime. *Physical Review Letters*, **101**, 011602.

Gleiser, M. & Sicilia, D. (2009). General theory of oscillon dynamics. *Physical Review D*, **80**, 125037.

Gleiser, M. & Sornborger, A. (2000). Long-lived localized configurations in small lattices: application to oscillons. *Physical Review E*, **62**, 1368.

Gleiser, M. & Stamatopoulos, N. (2012a). Entropic measure for localized energy configurations: kinks, bounces, and bubbles. *Physics Letters B*, **713**, 304.

Gleiser, M. & Stamatopoulos, N. (2012b). Information content of spontaneous symmetry breaking. *Physical Review D*, **86**, 045004.

Gleiser, M. & Thorarinson, J. (2007). A phase transition in U(1) configuration space: oscillons as remnants of vortex–antivortex annihilation. *Physical Review D*, **76**, 041701(R).

Gleiser, M. & Thorarinson, J. (2009a). A class of nonperturbative configurations in Abelian-Higgs models: complexity from dynamical symmetry breaking. *Physical Review D*, **79**, 025016.

Gleiser, M. & Walker, S. I. (2009b). Toward homochiral protocells in noncatalytic peptide systems. *Origins of Life and Evolution of Biospheres*, **39**, 479.

Graham, N. & Stamatopoulos, N. (2006). Unnatural oscillon lifetimes in an expanding background. *Physics Letters B*, **639**, 541.

Gunton, J. D. (1999). *Journal of Statistical Physics*, **95**, 903.

Gunton, J. D., San Miguel, M., & Sahni, P. S. (1983). The dynamic of first-order phase transitions. In C. Domb & J. L. Lebowitz (eds.), *Phase Transitions and Critical Phenomena*, vol. 8. London: Academic Press.

Hertzberg, M. P. (2010). Quantum radiation of oscillons. *Physical Review D*, **82**, 045022.

Hindmarsh, M. & Salmi, P. (2006). Numerical investigations of oscillons in 2 dimensions. *Physical Review D*, **74**, 105005.

Hindmarsh, M. & Salmi, P. (2008). Oscillons and domain walls. *Physical Review, D*, **77**, 105025.

Honda, E. & Choptuik, M. (2002). Fine structure of oscillons in the spherically-symmetric phi^4 Klein–Gordon model. *Physical Review D*, **68**, 084037.

Infeld, E. & Rowlands G. (2000). *Nonlinear Waves, Solitons and Chaos*. Cambridge, UK: Cambridge University Press.

Jeong, S.-O. & Moon, H.-T. (1999). Nucleation of oscillons. *Physical Review E*, **59**, 850.

Kauffman, F. (1993). *The Origins of Order: Self-Organization and Selection in Evolution*. Oxford, UK: Oxford University Press.

Landau, L. D. & Lifshitz, E. M. (1976). *Mechanics*. Oxford, UK: Pergamon Press.

Langer, J. S. (1992). *An Introduction to the Kinetics of First Order Phase Transitions*. In C. Godrèche (ed.), *Solids Far from Equilibrium*. Cambridge, UK: Cambridge University Press.

Lee, T. D. & Pang, Y. (1992). Nontopological solitons. *Physics Reports*, **221**, 251.

Lineweaver, C. H. & Egan, C. A. (2008). Life, gravity, and the second law of thermodynamics. *Physics of Life Reviews*, **5**, 225.

Melo, F., Umbanhowar, P. & Swinney, H. (1994). Transition to parametric wave patterns in a vertically oscillated granular layer. *Physical Review Letters*, **72**, 172.

Prigogine, I. (1978). Time, structure, and fluctuations. *Science*, **201**, 777.

Prigogine, I. (1980). *From Being to Becoming*. New York: WH Freeman.

Rajamaran, R. (1987). *Solitons and Instantons*. Amsterdam: North-Holland.

Scott, A. C. (2007). *The Nonlinear Universe: Chaos, Emergence, Life*. Berlin: Springer-Verlag.

Shats, M., Xia, H. & Punzmann, H. (2012). Parametrically excited water surface ripples as ensembles of solitons. *Physical Review Letters*, **108**, 034502.

Stenflo, L. & Yu, M. Y. (2007). Oscillons and standing wave patterns. *Physica Scripta*, **76** C1.

Tsimring, L. S. & Aranson, I. S. (1997). Localized and cellular patterns in a vibrated granular layer. *Physical Review Letters*, **79**, 213.

Umbanhowar, P., Melo, F. & Swinney, H. (1996). Localized excitations in a vertically vibrated granular layer. *Nature*, **382**, 793.

Umurhan, O. M., Tao, L., & Spiegel, E. A. (1998). Stellar oscillons. *Annals of New York Academy of Sciences*, **867**, 298.

Walgraef, D. (1997) *Spatiotemporal Pattern Formation*. New York: Springer.

Part III   Biological complexity, evolution, and information

7   Life: the final frontier for complexity?

Simon Conway Morris

The concept of complexity reminds one of the tasting notes of a rare vintage: everybody knows what you are talking about, but the realities continuously slip through our fingers. Moreover, in the scale of complexities most would agree that life is intrinsically more complex than, say, a galaxy. So too we suppose that some sort of metric stretches through the history of life: be it in terms of ecologies, bodyplans or nervous systems. In other words what we see today is manifestly more complex than what was found in the Precambrian. Yet an evolutionary perspective on complexity reveals some unexpected angles. To start with, although the history of life might fall into the cliché of "Once there were bacteria, now there is New York", in fact when one investigates what are evidently the most primitive representatives of a given group repeatedly they turn out to be "unexpectedly" complex. Many such examples are now available, but amongst the most telling are the eukaryotes. Second, there is the phenomenon of evolutionary inherency, the observation that much that will be required for the emergence of a complex form has already evolved at a substantially earlier stage. A good example involves the protein collagen, essential as a structural molecule in metazoans, but whose origins not only lie deeper in eukaryotic history but whose functions were evidently quite different. Inherency indicates, therefore, that much of complexity is nascent, almost homunculus-like, lying far deeper in the Tree of Life than generally appreciated. Third, whilst the arrow of time seems to lead to ever greater levels of organic complexity, it is as well to remember that these may well include

Complexity and the Arrow of Time, ed. Charles H. Lineweaver, Paul C. W. Davies and Michael Ruse. Published by Cambridge University Press.
© Cambridge University Press 2013.

examples that are often dismissed as "simplification" or "regression". Many of these instances, such as the dicyemid/chromodinid association with cephalopod renal organs or as an alternative example the inhabitants of an insect bacteriome, are intimately symbiotic and arguably are as complex as more familiar systems. Finally, I present evidence that biological systems are near, and in some cases have reached, the limits of complexity. Such is evident from areas as diverse as extremophiles, nervous and sensory systems, enzymes such as *Rubisco*, complex symbioses and functional complexes such as teeth and arthropod tagmosis. The paradox is that although evolution has run out of things to do, in the case of knowledge and wisdom the exact reverse appears to be the case. Here we have not reached the end of complexity.

Once there were bacteria, now there is New York. Those of a broader turn of mind might wish to substitute quarks – or some yet more fundamental entity – for bacteria, but by whatever register one chooses few would disagree that over time things have become more complex. What else can we say? Reporting as I do from a society that is obsessed with Health & Safety (the obverse of one possessing a moral compass, and thus where free will and responsibility for one's own actions are not subject to a legislation whose default position is that no risk is too small to avoid ludicrous and/or punitive consequences), it is only right and proper I provide a Risk Assessment. So let the reader be warned; as will become painfully apparent, I am no expert on the area of complexity. Perhaps this is not a fatal difficulty; few doubt that however important the topic may be it is frustratingly difficult to pin it down (Conway Morris, 2011c). So Risk Assessment in hand, please treat this chapter as a *tour de horizon*, short on general principles, let alone definitions, but one whose central thrust paradoxically is to look at the limits of what might be possible.

7.1 GRAPPLING WITH COMPLEXITY

In his on-line article Quantifying Complexity Theory (www.calresco. org/lucas/quantify.htm; accessed 13 July 2012) Chris Lucas provides a valuable over-view, reminding us that as in the past with figures such as Newton, Poincarě, and Cantor, so in complexity new mathematical techniques may need to be conjured up. So, too, he performs a valuable service in recognizing four types of complexity (static, dynamic, evolving, self-organizing), as well as outlining the sort of techniques (e.g. phase transitions, attractors) that might provide us with some sort of traction. It is no criticism that at least some of these lines of enquiry tend to be non-specific, show limitations of operation, or lead to an indeterminate number of outcomes. As Lucas also notes, too often in order to simplify a problem a reductionist element or tendency enters the picture. As he writes "This is rather strange, especially as the main problems we wish to solve are the prediction of complex overall structure (holistic view) and the analysis of systems with many simultaneous aspects (e.g. man)". Significantly, he immediately continues "Techniques to do either of these are still pretty well non-existent".

In Lucas' formulation of four types, the closest match to my aims would seem to be Type 3, that is evolving complexity. His thoughts in this area are interesting in as much as he remarks how the identification of similarities and their incorporation into categories is all very well and can "give useful general guidelines, but fail to provide one significant requirement for scientific work, and that is predictability". Whilst it might lack a theoretical framework, it is worth observing that some guide to predictability is surely available given the ubiquity of evolutionary convergence (Conway Morris, 2003, 2010a, 2010b, 2011a). This is important given Lucas' associated observations to the effect that "in any complex system many combinations of the parts are possible, so many in fact that we can show that most combinations have not yet occurred even once, during the entire history of the universe". This suggests that whilst in principle the combinatorial size of all biological possibilities is immense, the reality revealed by evolutionary convergence is that some, arguably most, combinations have already been tried out and not found wanting.

This will, of course, seem a surprising claim, one quite at odds with the neo-Darwinian perspective that insists (for metaphysical reasons) on the indeterminacy of outcomes. Space does not allow any proper unpacking of why this perspective may be wide of the mark. Or is it? After all, it would be ludicrous to dispute that biology is haunted by Elsasser's Immense Numbers (e.g. Elsasser, 1998), with the combinatorics so stupendously large that the last four billion years of terrestrial evolution (or c. 10 billion years if we assume life is restricted to Earth-like planets; see also Setiawan *et al.*, 2012) would only allow the exploration of an infinitesimally small fraction of biological hyperspace. This presupposes, however, any given combination is equally likely to emerge, survive, let alone propagate. This, of course, is extremely unlikely. Be it in terms of biophilic amino acids (as well as their chirality), the genetic code, or enzymatic convergences, and that is just the start of the story, it can be argued that the substrate of possibilities that are pre-determined by the physico-chemical conditions of the universe will ensure that practically all of biological hyperspace will remain unvisited, not because there was insufficient time but because the hypothetical alternatives will never actually work. If that proves correct, then not only is life extremely finely poised between non-viable states that are quasi-crystalline or chaotic (see Macklem, 2008), but evolution itself is forced to navigate the silver threads of vitality that define the handful of routes across an otherwise vast and dead landscape. If, as I suggest, life is as fine-tuned as the rest of the universe, then the limits to complexity may be easier to discern.

To revert to my Risk Assessment: any claim I have to competence in this chapter is as a biologist. My intention is to outline an approach to biological complexity that asks not how we might define it but whether there are genuine limits. If so, then perhaps with fences, hedges, and barriers we can pin down the beast (Conway Morris, 2011c). Much of what I can say has arisen from my readings around the topic of evolutionary convergence. So far as I am aware it has not been set out in this fashion. To that extent only could it be claimed to be new.

7.2 WHICH METRIC?

It is not, however, the only approach, and by way of general introduction it is worth touching on some other approaches. Foremost in this regard, perhaps, is the work by Dan McShea (e.g. McShea, 1998, 2002), and points of congruence might be found between his series of penetrating insights and the observations provided in this chapter. Here, however, I touch briefly on one popular metric. This is the number of different cell types in animals, a yardstick that at first sight seems to offer a useful way forward (e.g. Valentine, 2000). After all, if sponges have c.15 cell types and humans some fifteen times more (c.215), then do we not have something concrete to talk about? Matters, however, might not be quite so simple. This is not to contest the relative simplicity of a sponge as against a human, but to draw attention to some additional considerations.

First, we need to be careful not to under-estimate the complexity of some "primitive" organisms. In this regard the benthic planula of a hydrozoan (Cnidaria), known as *Clava multicornis*, provides a useful object lesson. This is because it displays a "surprisingly" complex behavior that is underpinned by both a striking polarization of its nervous system and array of sensory cells and also an "unexpected" degree of neural organization and cellular diversity (Piraino *et al.*, 2011). In this context it matters little if this quasi-bilaterian organization arose independently or drew on the genomic resources of a common ancestor. This is because when it comes to assessing relative complexity it is important to realize that in morphologically "simple" groups such as the sponges (Harcet *et al.*, 2010; Srivastava *et al.*, 2010) and cnidarians (Putnam *et al.*, 2007) the relatively limited range of cell types is in dramatic contrast to their degree of genomic complexity. This is not to deny, and especially with respect to the sponges, that in due course the evolution of additional genes and rounds of gene duplication will be necessary for further phenotypic elaboration, but so too there is evidence that genomic histories (especially in the ecdysozoans, e.g. fly, worm) are complex and involve significant gene loss. What is even more important is, as Harcet *et al.* (2010) note, that with respect to the cnidarians, or at least the model organism the sea-anemone *Nematostella*, "much of the genome complexity in terms of gene content and structure was already present in the common ancestor of all Eumetazoa" (p. 2747). This includes the nervous system, important components of which have evolved not only amongst the sponges (Nickel, 2010; Srivastava *et al.*, 2010) but even deeper amongst pre-metazoans (e.g. Burkhardt *et al.*, 2011; Ryan & Grant, 2009). The significance is not that any such "primitive" organism is more complex than usually thought, although it should now be apparent that customary metrics for complexity may run into ambiguities, but that inherent in the molecular constitution of sponges and the like are the potentialities for more complex systems. From this perspective the emergences of complex nervous systems, as indeed foreshadowed by the hydrozoan planula (Piraino *et al.*, 2011), are very probable, if not inevitable.

7.3 LET THERE BE BRAINS

So far as nervous systems are concerned an end-point is the human brain. So, too, it is commonplace to suggest that this is the most complex structure in the universe. If we ever find alien life, then this statement may need to be revised, but taking our brains as the yard-stick it is still worth remembering that this specific path to complexity really began c. 1.5 Byr ago. Why then? Because this is the datum point whereby the recognition of that genetic component of yeast and plants that subsequently we find employed in the nervous systems of animals (Mineta *et al.*, 2003) can be taken as the first point in time when a future nervous system becomes likely, if not a strong possibility. Note also that Mineta *et al.* do not simply identify some generalized suite of proteins in yeast and plants, but ones that subsequently find specific employment in various neural categories (such as neurotransmission, neural differentiation, etc.). Nervous systems themselves are unlikely to be much older than about 580 Myr (Pecoits *et al.*, 2012), brains (of a sort) followed soon after (and certainly by 530 Ma). Subsequent to this in the vertebrates at least there were various increases in brain size, but rampant encephalizations were, it seems, effectively only initiated in approximately the last 20 Myr (current data point to c.18 Myr for dolphins (e.g. Marino *et al.*, 2004), c.7 Myr for hominids (e.g. Zollikofer *et al.*, 2005) and perhaps much the same figure for the New Caledonian crows (e.g. Haring *et al.*, 2012; see also Cnotka *et al.*, 2008)).

Irrespective of how precise these figures are, they simply remind us that from one perspective, that of nervous systems, complexity was a long time a-coming, but when it did the pace picked up, perhaps exponentially and much more intriguingly with some degree of synchronicity. One could, of course, choose a myriad of other examples from biology, and it would be a mistake to assume that matters proceed to some sort of agreed time-table. Photosynthesis, for example, almost certainly dates back to at least 3.5 Byr, and apart from carbon concentrating mechanisms, of which the geologically recent rise of convergent C_4 photosynthesis (Sage, 2004) is the best-known, it is not obvious that the mechanisms of photosynthesis can have improved much since the Archaean. But the increase in the complexity of the nervous system has a peculiar relevance, because it is only (again so far as we know) by this agency that the universe has become self-aware and at least one of its representatives decided that complexity is an issue worth addressing.

What then is worth saying, other than 3.5 Byr down the line we have New York (or Bach's *B Minor Mass*, Giorgione's *Tempest*, or whatever one's preferred metric might be)? Here I will introduce four topics to suggest that at least so far as biology is concerned the concept of complexity is – well – complex. The specific issues to be addressed are: (a) how simple are the starting points, (b) what is inherent in the process, (c) is complexity irreversible, and (d) are there ultimate limits to biological complexity? In all these cases I suggest the answers are not what current neo-Darwinians might think. Nor, in parentheses, should this apparent aside be regarded as a swipe at evolution. The facts of evolution are not in dispute, but we should not be shy of attempting to put them in a wider framework. Most obvious in this respect is self-organization (e.g. Salthe, 2008), which in many ways has been the ghost at the Darwinian banquet since at least the time of D'Arcy Wentworth Thompson (1942). Nobody can doubt the importance of self-organization, but despite all the elegance of life's geometries (be they logarithmic spirals, Fibonacci series, helical arrangements, etc.) the subject still has an embarrassment of promissory notes. No science of biological complexity can possibly be complete without reference to self-organization, but here I will take a series of more empirical observations to suggest complexity has envelopes of possibility, with one paradoxical exception – ourselves.

7.4 HOW SIMPLE ARE THE STARTING POINTS?

"Mind the gap" as they say on the London Underground. And such an announcement should resonate with anybody working not only on the origin of life, but any other step-change in complexity during the evolution of life. To be sure the origin of life might be the most glaring example, but topics as diverse as explaining the origin of eukaryotes or language are notoriously refractory. Might there, however, be some common principles? Rather than beating our head against the brick wall as to how we bridge the gap between a prebiotic porridge of not-very-exciting molecules into a functioning cell with perhaps 400 genes, we might do better to focus on examples where some progress may be feasible. The first paradox to note is that it may be that things were never very simple to start with. In other words, whilst the "Once there were bacteria, now there is New York" broadly holds, it may be a mistake to suppose that at their initiation the object in question was grovellingly simple.

Two lines of evidence support this conjecture. One, that of evolutionary inherency, already hinted at with respect to the origin of the nervous system, will be returned to below. The second revolves around the under-appreciated fact that some instances are far from the sort of scenario where a given novel evolutionary capacity stumbles into the sunlight with hardly a loin-cloth to its name; in fact it has already acquired a more than passable sartorial elegance. Take the early history of eukaryotes. Why the eukaryotes? Would it not be more apposite to start as near "the beginning" as we can? Call it cowardice if you like, but I call on the eukaryotes for the following reasons. Firstly, it appears that a good deal of what one needs to make a eukaryote is inherent in the prokaryotes (e.g. Graumann, 2007). Secondly, inter-relationships amongst the bacteria are not fully resolved and horizontal gene transfer is very common. In other words their story is more complex, but perhaps in a different sort of way. Thirdly, and most importantly, although types of multicellularity and organelle development (e.g. Keim *et al.*, 2004) are more widespread amongst prokaryotes than often appreciated, it is within the eukaryotes that we look to features such as tissue differentiation, cell-signaling, vesicle formation, membrane fusion, etc, all of which are prerequisites for complexity.

Knowledge of the eukaryotic tree and reasonably extensive sampling of their genomes allows one to infer the nature of the "first eukaryote", that is the genetic tool-kit that has been inherited by all descendant forms (at least in principle). Slobberingly simple? Gratifyingly primitive? Agreeably archaic? Not at all. In case after case as the molecular data-bases grow it transpires that far from being equipped with some sort of rudimentary tool-kit, as far back as one can peer the given repertoire is already quite complex. Best known, perhaps, are the examples of the so-called SNAREs (an abbreviation for the proteins known as N-ethylmalemide-sensitive factor attachment protein receptors), which play a key role in the evolution of multicellularity and cellular complexity (e.g. Kloepper *et al.*, 2007, 2008). To be sure, in the course of a myriad of evolutionary trajectories these capacities depended on the expansion of the SNARE family, typically by gene duplication, but the point is that evidence now shows how even the ancestral eukaryotes must have possessed an elaborate system.

And as the evidence grows much the same applies to other complex molecular systems such as the proteins associated with a structure (the midbody) that is important in cell division. So, too, with respect to the genetic machinery, a survey of eukaryotic genomics shows that even "primitive" eukaryotes had complex genetic machinery. Thus Fritz-Laylin *et al.* (2010) remark how their study of an amoeboflagellate "reveals [an] unexpectedly rich versatility [of such machinery] in early eukaryotic ancestors" (p. 639).

To this writer words like "unexpected" (or "surprising") are like the sound of tapping from sappers working deep beneath supposedly secure fortifications. How then are we to explain the complex nature of ancestral forms? One could, of course, argue that the real ancestors have long since been obliterated, taking with them any evidence of their rudimentary nature. Alternatively, one could look to rampant horizontal gene transfer that has blurred the evolutionary picture beyond recognition. Such are indeed pitfalls, and were it but one or a few instances of "unexpected" complexity in the ancestral state then one would need to exercise extreme caution. But the reverse is evidently the case. Thus, in the case of early eukaryotes, along with the examples given above, one can add others. One such involves the homeodomain proteins, which fall into two classes, TALE and non-TALE (TALE being an abbreviation for *Three Amino Acid Loop Extension*, and refers to the three amino acids that are inserted into the protein and link helices 1 and 2; see Bertolino *et al.*, 1995). Again the evidence suggests that their original configuration was complex and where absent this is due to secondary loss rather than a primitive simplicity (Derelle *et al.*, 2007). Homeodomains, of course, are ultimately central to the construction of bodyplans, but at the cellular level another class of proteins, the kinesins, provide molecular motors that are essential for the cytoskeleton. Of course, like the other examples, kinesins have shown impressive diversifications during eukaryotic history. Nevertheless, as Wickstead *et al.* (2010) observe, the early eukaryotes "already possessed a highly complex cellular form *before* giving rise to any of the sampled extant eukaryotic groups [Wickstead *et al.* looked at five of the six extant major groups]. This proto-eukaryotic cell was surprisingly highly developed in terms of kinesin motor types" (p. 8; their emphasis).

Again note the word "surprisingly". These examples of complexity at the dawn of eukaryotic history surely point to the capacity for evolution not only to self-assemble the necessary molecular machinery but to ensure that the component parts are integrated. This may sound like a covert plea for "Intelligent Design". It is nothing of the sort, not least because there is no intention to employ some fairy-tale like notion such as "irreducible complexity". What these examples do suggest, however, is evolution is governed by rules of organization that do not themselves depend on protein or cellular chemistry. But that is not the end of the story.

7.5 EVOLUTIONARY INHERENCY

It is commonplace that evolution has no fore-knowledge. This is correct, but it overlooks the fact that, given the emergence of a given structure, in many cases a good argument can be made that a certain level of complexity will then be highly probable, if not inevitable. Writing of the homeodomain proteins, Derelle *et al.* (2007) argue that their early occurrence indicates "that the eukaryotes as a whole are adapted to multicellularity" (p. 217), and much the same can be inferred from the SNAREs. In one respect this molecular inherency is hardly exceptional, inasmuch as it can be regarded as more or less synonymous with the evolutionary principle of co-option whereby a component that has evolved in one context is employed (or, if you prefer, hi-jacked) in an entirely unrelated one. The crystallins of eyes are perhaps the best-known example, not only on account of their convergent employment but because the battery of proteins that have been co-opted in a myriad of different animal eyes originally had entirely different functions in microbes, often related to stress control. Co-option and inherency are, however, not quite the opposite sides of the same coin. Mention the word "co-option" and most evolutionary biologists will sagely nod their heads. Yet, its ubiquity is under-appreciated.

Take, for example, the curious case of collagen, without which animals could not exist given its central role as a structural protein. Passing by the intriguing examples of its convergent evolution in viruses (Li et al., 2004), bacteria (Waller et al., 2005) and fungi (Wang & St. Leger, 2006), one might expect collagen to have evolved at the dawn of animals. Some proteins, such as those of the Hox family, evidently have. But not collagen, because this occurs in the choanoflagellates which are the sister-group of the animals. Yet the specific choanoflagellate, the taxon Monosiga, is single-celled and clearly its collagen has no structural role (King et al., 2008). So what does it do? There is a hint that initially it was employed in cell-signaling (Heino, 2007). But co-option for such a structural role means that it is not ridiculous to argue that our Achilles' tendon was inherent in choanoflagellates. And this is where the distinction between co-option and inherency lies. This is because the latter subsumes the former, but also has the implication of high probability or inevitability. In other words, complex forms cannot help but arise not only because there is a "landscape" across which evolution is compelled to navigate, but because many of the component parts have already evolved.

In passing it is worth noting these notions of inherency are by no means torpedoed by the concept of so-called "deep homology". I have addressed these issues elsewhere (Conway Morris, 2010a), but in brief the idea is that because we see many examples of genomic conservation in otherwise only distantly related animals, features as diverse as excretory systems, brains and such neural components as the mushroom bodies of some protostomes, and sleep are regarded as fundamentally homologous. Perhaps the classic example of supposed deep homology is the case of eye development which by and large depends on the Pax-6 gene. The argument then goes that not only are all eyes effectively homologous but if Pax-6 did not exist then neither would eyes. Such has been seriously proposed (e.g. Gehring, 2005), yet in my view it spectacularly misses the point. If it were correct, of course, then many roads to complexity might be irreversibly de-railed. The reason not to be perturbed by this scenario is briefly as follows.

Are not eyes inherent from the dawn of animals? Indeed they are, but this has little to do per se with *Pax-6*. First, of course, the opsins (part of the enormous class of *G*-coupled protein receptors) and crystallins are recruited as part of the evolution of eyes. In some cases they might be regulated by *Pax-6*, but they are neither coded for by nor rely on this gene. Second, the evolution of an eye may indeed require a paired-box (i.e. *Pax*) gene in as much as any eye has two key components, that is a pigment for shielding incoming light from given directions and an opsin (Kozmik, 2008). This, however, is a functional constraint, and recalls other evidence for genetic predisposition for at least some developmental changes (e.g. Christin *et al.*, 2010). What then specifically of *Pax-6*? Firstly, this gene did not evolve specifically in the context of the evolution of eyes. Its original functions, so far as can be discerned, were involved with the expression of anterior nervous tissue. So it is unsurprising that *Pax-6* is expressed not only in the eye, but also the nose and parts of the brain. Secondly, *Pax-6* is not primitive to animals per se, and in the case of the cubozoans (a group of jellyfish, so relatives of the corals) not only do they possess convergent camera eyes, but the *Pax* gene they employ is not directly related to *Pax-6* (see Kozmik *et al.*, 2003, 2008; specifically in the cubozoans the eye is regulated by *Pax-B*, whereas it is the co-occurring *Pax-A* that is evolutionarily closer to *Pax-6*). To be sure, with few exceptions in more advanced animals eye development depends on *Pax-6* (or its equivalents). Here, as elsewhere, there may be a conservation of development but importantly the deployment and timing of the genetic cascades are by no means the same (see also Royo *et al.*, 2011). Thirdly, *Pax-6* is also redeployed, perhaps most famously in the case of bird muscle (Heanue *et al.*, 1999). Such redeployment is, of course, the evolutionary norm, but it is a reminder that there is seldom (if ever) a one-to-one relationship between a developmental gene and a structure. Finally, and in a case we will return to below, in highly simplified metazoans, specifically the dicyemid mesozoans, the animal has not only lost its eyes (and there are reasonable grounds to think its ancestor possessed eyes) but its entire nervous system. Yet mesozoans still retain *Pax-6* (Aruga *et al.*, 2007). Quite why seems to be obscure, but one can be sure that not only is it functional but engaged in some vital role such as calotte formation.

All this shows is that complex anatomical structures require developmental genes, and given that the general class of homeotic genes is evolutionarily very ancient (Kant *et al.*, 2002), then it will be unsurprising that they not only evolve but are repeatedly redeployed. More generally, as with the term co-option, there remains what I believe is an important distinction between what I term inherency and the concept of deep homology. The former invokes potentiality, with the suggestion that much of what might be regarded as complex is seeded within more primitive organisms. To be sure, their emergence needs time and the ability for once-separate genetic systems to interact (as often as not by the evolution of new genes). Deep homology, on the other hand, is much more literalistic. In my view the concept tends towards an essentialism that falls into the trap of "genes for something", exemplified by *Pax-6* in eye development (e.g. Gehring, 2005). This provides an over-reductive framework and, although it might hint at pre-requisites, it leaves unexamined both how such systems can be so readily redeployed and what the dynamics of each evolutionary pathway are. This distinction, I suggest, is far from academic. Deep homology more or less takes a given gene as an evolutionary fact, a convenient monad, whereas inherency is far more ambitious in enquiring as to whether evolution might depend upon particular organizational principles.

7.6 REVERSING COMPLEXITY

The mention of the mesozoans leads to another aspect of evolutionary biology that is familiar enough to its practitioners, but perhaps neglected by those who see evolution as principally exploring ever more complex domains. Dicyemid mesozoans are remarkable animals, that are only found in association with the kidneys of cephalopods (Furuya & Tsuneki, 2003). They are vermiform and spectacularly simplified, consisting of less than 100 cells. In passing we should note they show an intriguing convergence with an unrelated group, the chromodinid ciliates which also inhabit the surfaces of cephalopod kidneys (Hochberg, 1982). One could observe that this very specific association of vermiform creatures with the cephalopods is by no means accidental but is actually an essential ingredient in the success of this group, specifically enhancing the function of the kidney. As is well known, cephalopods show many striking convergences with the vertebrates, and the case of the mesozoan/chromodinid association might contribute to our understanding as to how a relative of the limpet achieves a level of complexity that seems to test the design capabilities of the molluscan bodyplan to the limits of the possible (Conway Morris, 2011a). But to return to the dicyemids, and in passing to note that these are not the only example of a dramatic simplification amongst the metazoans (consider the myxozoans, e.g. Kent *et al.*, 2001), in the context of this chapter their importance is to recall that the reverse of complexity is always on the cards. Or is it really a reverse? Recall first their specific location on the surface of a cephalopod kidney, drenched in urine. With few exceptions the cephalopods require the mesozoan (or effectively equivalent chromidinid ciliate), and as argued they may be essential to such a complex animal. And, correspondingly, although long regarded as parasitic, this is evidently a mutually beneficial association. Indeed one could argue that it is a mistake to separate "host" and "guest"; they are utterly interdependent and the complexity of the former would not be possible without the latter.

Very similar arguments, of course, can be applied to other symbiotic associations, perhaps most obviously the role of γ-proteobacteria like *Buchnera* in the groups such as the hemipteran insects whereby an otherwise problematic food source (nectar, xylem, etc.) is rendered accessible. In this latter case, which unsurprisingly provides some excellent examples of convergent evolution, the associations are evidently mutually beneficial.

At first sight it would be difficult to argue that this applies when one comes to the various pathogenic bacteria. Here, too, there are intriguing examples of convergence, whereby massive genomic loss (as indeed we also see in the insect endosymbionts) is a result of the co-option of the host's genetic machinery. But again, as a system is it any less complex? One might argue that whilst the convergences remind us of the restriction of possibilities, both the loss of the genome and the co-option of the host's genetic machinery represent an integration of function of a very high order. It is not unusual for such organisms to be labelled as degenerate, regressive, or some similar epithet. Such a terminology misses the point and the interlocking of genomes seems to point towards an under-appreciated degree of complexity. It may be no accident that the estimate of a minimal genome size of any cell is based on the parasitic bacterium *Mycoplasma* (Fraser *et al.*, 1995), with the implication that such an organism has reached a boundary of the biological universe. It might at first be thought that so far as complexity in the opposite direction is concerned, that is away from the purportedly "simple" and to the realm of multicellular organisms, then so far as such boundaries are concerned they are effectively non-existent. This may not be the case.

7.7 ARE THERE LIMITS TO BIOLOGICAL COMPLEXITY?

The Darwinian mantra broadly holds that, other than the exigencies of the death of Solar Systems (and yet more distant catastrophes) evolution is both indeterminate and open-ended, with the widespread view that it is unpredictable and the implicit (albeit largely un-examined) assumption that there is no limit to its complexity. The case against its indeterminacy I have addressed elsewhere (Conway Morris, 2003, 2010a, b), but can we make the case that life has already begun to reach the boundaries of the biological universe (however defined)? The implication would then be that any room for exploring complexity much further is now highly restricted. Because we lack a description of biological hyperspace it is very difficult to offer more than a few pointers that depend on widely disparate examples. Yet, because at least some of these examples, such as the constraints of nervous systems, depend on general factors such as allometry and energy consumption, we have some confidence that our conclusions are not entirely ham-strung by merely parochial circumstances.

Consider first the physical and chemical limits of life. The organisms that inhabit environments that to us would be toxic, scalding, or equally lethal are labelled the extremophiles: our view, not theirs. On the perhaps not unreasonable assumption that many exoplanets even in the habitable zones of their host stars are inclement, extremophiles are naturally a focus of astrobiological interest. To begin with, it is evident that many of the solutions, be they living in near-boiling temperatures or in conditions of extraordinary alkalinity or salinity, evolved multiple times (e.g. Mongodin *et al.*, 2005; Puigbò *et al.*, 2008). This in itself suggests that occupation of such zones is unlikely to be simply the result of highly fortuitous evolutionary events. Less appreciated, however, is the fact that the limitations that terrestrial extremophiles have reached – be it in terms of temperature, pressure, salinity, pH, or water activity – seem to be close to those capable of carbon-based life forms anywhere (Conway Morris, 2010c, 2011b). So what has this to do with complexity? Two things. First it is a reminder that the envelope of biological possibilities is real, and that these alone constrain what is possible. Second, it is likely that the molecular solutions to deal with a given extreme may be adaptively spectacular (fancy drinking hot battery acid?) but they are often subtle in comparison with their mesophilic relatives and most likely revolve around largely unappreciated sensitivities.

This combination of convergence and molecular fine-tuning suggests that extremophiles are a tractable area to study complexity. How so; is not this claim indicative of how difficult it is to tie down concepts of complexity? Maybe so; obviously the metric for such extremophiles is going to be difficult to apply to the discussion, given above in Section 7.2, concerning the number of cell types in animals (and all its attendant difficulties). Rather, and perhaps to reiterate, it is the proposal that the envelope of physico-chemical possibilities for life are real: bacteria will *never* live at 150°C or in an environment where the water activity is less than 0.6. Perhaps "out there" or in some unexamined region of our planet they can, but if it transpires that the envelope remains unbreached then this limit surely tells us something important about the universal constraints on what life can, and more importantly, cannot do. One might observe, of course, that bacteria do other things, sometimes of seemingly surprising complexity. Consider their social behavior. The point here is first to stress that not only are they complex but they show some striking convergences with eukaryotic systems (Crespi, 2001). Next is to enquire whether in this area prokaryotes and/or eukaryotes have reached the limits of what is possible and, if so, whether there are unexpected similarities in their otherwise disparate systems.

If the physico-chemical boundaries of life are already defined, what of the denizens of the more benign regions where the vast majority of species actually live? Again the Darwinian mantra is that what we see is good enough in its own way, no doubt capable of improvement should the adaptive circumstances demand, but in the grand scheme of things far from perfect. To be honest, a bodge-job, impressive enough, but the rivets showing everywhere. The evidence actually suggests otherwise, and in doing so indicates that, with one crucial exception, not only is there a limit to biological complexity but in a way analogous to the extremophiles these limits may be close to being reached. The evidence in support of this radical conclusion cannot be summarized here other than by reference to a few examples. More importantly, such evidence is by no means complete. Rather, my purpose, as with the earlier argument that ancestral forms are far from slobbering balls of simple slime, is to propose a research program that looks beyond the current neo-Darwinian framework. Not, of course, to deny its many insights but simply to enquire whether it provides a total explanation. It seems otiose to remind the reader, but perhaps is still necessary, that, contrary to the still fashionable methodology of cladistics that treats organisms as atomistic, not even as an "exploded" version of a machine but a barrel of bits and pieces, the reality is that any organism is functionally highly integrated. This in itself is a reminder that, even if a given structure looks sub-optimal, in reality it is still possible that it is as good as it ever will be. In certain cases authors are un-embarrassed to argue that their given examples are not just fit for purpose, but verge on the optimal. "The tooth of perfection" as Evans and Sanson (2003) remark in the context of the evolution of cutting surfaces, and this finds a corresponding echo in the identification by Kuch *et al.* (2006) of snake fangs being effectively perfect. In the latter case, unchanged as they have been for at least the last 20 Myr, this suggests that there is little if any scope for further improvement.

Such examples, of course, resonate with human aspirations: we may indeed muddle along but even a moment's detour to the arcane area of sport reveals a craving for the competitors to exceed the limits. Correspondingly, walking disasters may excite our pity, but not our admiration. But perhaps, at least in the biological sphere, we should be a little bit more open-minded. Take the protein *Rubisco*, of central importance in photosynthesis where as a carboxylase it acts as the enzyme central to the fixation of carbon. Yet, as is well-known, its rate of activity is cripplingly slow, explaining why it is the most abundant protein on the planet (Ellis, 1979). It is difficult to think of an example where even a modest improvement in efficiency would not be adaptively highly significant, yet the only solution appears to be by rampantly convergent methods of carbon dioxide concentration, be they by carboxysomes (e.g. Badger *et al.*, 2002) or specialized methods of photosynthesis (C$_4$, see Sage, 2004 or CAM, Lüttge, 2004). It may be that *Rubisco* is as good as it can get; as Tcherkez *et al.* (2006) note "all ribulose bisphosphate carboxylases may be near perfectly optimized". Such examples as *Rubisco* are a useful reminder of some under-appreciated realities: even what appears to be disastrously mal-adaptive – who, after all would design an enzyme that is disabled by the very substance (oxygen) it helps to produce? – may still be as good as is possible.

If the human brain is indeed the acme of evolutionary complexity, we can still doff our hats to the remarkable sophistication of the photosynthetic mechanisms, acknowledging that in photosynthesis and, if Roger Penrose (e.g. Penrose, 1994) is correct, also our brains, quantum processes are central to their operation. Certainly *Rubisco* seems to be the exception amongst enzymes, many of which (such as the convergently evolved carbonic anhydrase) seem to verge on the unbelievable in their efficiency. The main point, perhaps, is that it is all very well talking about biological complexity, but its emergence depends not only on the integration of form but quite possibly apparent compromises of function that paradoxically are still near-optimal.

As already indicated, the question of whether complexity is close to its limits at present can only be addressed on a case by case basis, but in any of these examples one sees how they might point to underlying general principles. Think, for example, of the case of agriculture in animals. This is convergent, and amongst the less familiar examples, perhaps, are the ambrosia beetles (e.g. Farrell *et al.*, 2001) and damselfish (e.g. Hata & Kato, 2006). But the canonical example is the leaf-cutter attine ants. Whilst the agricultural practices in the attines evolved about 50 Ma, the more sophisticated leaf-cutters only appeared about 12 Ma (Schultz & Brady, 2008). As is well known, this is an extremely sophisticated system, involving, amongst other things, cutting of leaf material, transport to the nest, preparatory cleaning, employment of minims and in some species particular task assignments, preparation of a substrate, maintenance of a monoculture of fungi, weeding, application of herbicides and other antibiotics, and waste disposal. Particularly remarkable is the employment of actinomycete bacteria, attached to the ant bodies, that produce the antibiotics. It is a complex system, but can it get any more complex? It is not clear it can. Such an example has, of course, a bearing on the many other examples of symbiosis. Whilst nobody doubts their complexity and capacity to integrate, it would not be so surprising if there are limits to the number of organisms that can be involved in a given association. Yet, given that the agriculture of the attine ants depends on just such associations, one might argue that they have gone as far as any social insect that decides to raise crops will ever get, here or anywhere.

As a final example, consider the case of complexity that revolves around the construction of a body-plan, and specifically the crustaceans. As is well known, this group shows an impressive ecological range, including the robber crab that apart from its larval stage is entirely terrestrial. So, too, they exemplify the principle of tagmosis, that is the dedication of particular parts of the body and especially the limbs to specific tasks that may revolve around walking, respiration, and the various complexities of feeding. The assumption is that from an effectively uniform series of appendages they have become successively specialized. In passing we should note that just such a primitive example, with minimal tagmosis, characterizes the remipedes. They were long assumed to be gratifyingly primitive, ready made to be pinned to the bottom of the crustacean tree. This, however, turns out to be almost certainly incorrect and so serves as a reminder of how the apparently simple might conceal remarkable complexity. At least so far as the brain is concerned, this is exactly the case in the remipedes (Fanenbruck & Harzsch, 2005). But let's return to the more familiar territory of tagmosis. As Adamowicz *et al.* (2008) demonstrate, not only has the trend of increasing tagmosis evolved independently a number of times (a hardly surprising conclusion), but in the context of this chapter more significantly, the limits of tagmosis seem to be close to being reached. But not quite. Intriguingly, as the authors point out, it is difficult to decide whether this is because this evolutionary potential remains untapped, or whether some other constraint (be it developmental, ecological, or functional) means that the last lap will always remain out of reach. As with the attine ants and their symbioses, so this example of complexity in the organization of a body-plan might point to similar cases that allow a more general principle to be derived.

But of all the limits to complexity perhaps the most interesting is the one that concerns the evolution of nervous systems. It is well known that in terms of metabolic energy nervous systems are cripplingly expensive: the retina of the blow-fly alone accounts for an extraordinary 8% of the total energy budget of the insect (Laughlin et al., 1998). So, too, there is an important literature on the various ways by which nervous systems can be employed to greatest effect with the minimum expenditure of energy. These, in their various ways, indicate that however complex nervous systems may get, there are definite limits to what in the end is possible. This is not to say that there is a single solution, and in this context one might observe that although encephalization is usually associated with tissue ther-mogenesis, this obviously does not apply to the octopus. Neverthe-less, when we look at the evolution of the mammalian brain then, as Hofmann (2001) has shown, there appear to be definite limits to any further increase in size and by implication its complexity. In part this revolves around the capacity of a brain that, ever increasing in size, still has to engage in effective integration, not least as the number of sensory fields mushrooms. Of equal importance is that the different allometries of the grey and white matter seem to put an absolute limit on the brain size such that it cannot increase in size by much more than three times (Hofmann, 2001).

So here, in a way, we reach the final level of complexity, at least at a biological level. We can't run at 40 mph (although a spurt of c. 23 mph is by no means negligible), nor fly (but a gin and tonic at 38 000 feet has its merits), nor swim the Pacific (although the free diving record of 800 feet deserves a salute), but we can out-think any other organism that has ever evolved. But have we, too, reached the limits of neural complexity and capacity to ask, let alone understand, the next set of questions? I suggest not, but that, as they say, is another story.

ACKNOWLEDGEMENTS

Warm thanks to Vivien Brown for typing numerous versions of this manuscript. I also thank Mary-Ann Meyers, Paul Davies, and the John Templeton Foundation for inviting me to the symposium on complexity in Phoenix, and the helpful comments and suggestions from an anonymous reviewer.

REFERENCES

Adamowicz, S. J., Purvis, A., & Wills, M.A. (2008). Increasing morphological complexity in multiple parallel lineages of the Crustacea. *Proc. Natl. Acad. Sci. USA*, **105**, 4786–4791.

Aruga, J., Odaka, Y. S., Kamiya, A., & Furuya, H. (2007). *Dicyema* Pax6 and Zic: tool-kit genes in a highly simplified bilaterian. *BMC Evol. Biol.*, **7**, art. 201.

Badger, M. R., Hanson, D., & Price, G. D. (2002). Evolution and diversity of CO_2 concentrating mechanisms in cyanobacteria. *Functional Plant Biol.*, **29**, 161–173.

Bertolino, E., Reimund, B., Wildt-Perinic, D., & Clerc, R. G. (1995). A novel homeobox protein which recognizes a TGT core and functionally interferes with a retinoid-responsive motif. *J. Biological Chem.*, **270**, 31178–31188.

Burkhardt, P., Stegmann, C. M., Cooper, B. *et al.* (2011). Primordial neurosecretory apparatus identified in the choanoflagellate *Monosiga brevicollis*. *Proc. Natl. Acad. Sci. USA*, **108**, 15264–15269.

Christin, P.-A., Weinreich, D. M., & Besnard, G. (2010). Causes and evolutionary significance of genetic convergence. *Trends Genetics*, **26**, 400–405.

Cnotka, J., Güntürkün, O., Rehkämper, G., Gray, R. D., & Hunt, G. R. (2008). Extraordinary large brains in tool-using New Caledonian crows *(Corvus moneduloides)*. *Neurosci. Letters*, **433**, 241–245.

Conway Morris, S. (2003). *Life's Solution: Inevitable Humans in a Lonely Universe*. Cambridge: Cambridge University Press.

Conway Morris, S. (2010a). The predictability of evolution: glimpses into a post-Darwinian world. *Naturwissenschaften*, **96**, 1313–1337.

Conway Morris, S. (2010b). Evolution: like any other science it is predictable. *Phil. Trans. R. Soc. Lond. B.*, **365**, 133–145.

Conway Morris, S. (2010c). Aliens at home? *EMBO Rep.*, **11**, 563.

Conway Morris, S. (2011a). Consider the octopus. *EMBO Rep.*, **12**, 182.

Conway Morris, S. (2011b). Predicting what extraterrestrials will be like: and preparing for the worst. *Phil. Trans. R. Soc. Lond. A.*, **369**, 555–571.

Conway Morris, S. (2011c). Complexity: the ultimate frontier? *EMBO Reports*, **12**, 481–482.

Crespi, B. J. (2001). The evolution of social behavior in microorganisms. *Trends Ecology Evol.*, **16**, 178–183.

Derelle, R., Lopez, P., Le Guyader, H., & Manuel, M. (2007). Homeodomain proteins belong to the ancestral molecular toolkit of eukaryotes. *Evol. Dev.*, **9**, 212–219.

Ellis, R. J. (1979). The most abundant protein in the world. *Trends Biochem. Sci.*, **4**, 241–244.

Elsasser, W. M. (1998). *Reflections on a Theory of Organisms: Holism in Biology.* Baltimore: John Hopkins University Press.

Evans, D. R. & Sanson, G. D. (2003). The tooth of perfection: functional and spatial constraints on mammalian tooth shape. *Biol. J. Linn. Soc.*, **78**, 173–191.

Fanenbruck, M. & Harzsch, S. (2005). A brain atlas of *Godzilliognomus frondosus* Yager, 1989 (Remipedia, Godzilliidae) and comparison with the brain of *Speleonectes tulumensis* Yager, 1987 (Remipedia, Speleonectidiae): implications for arthropod relationships. *Arthropod Struct. Dev.*, **34**, 343–378.

Farrell, B. D., Sequeira, A. S., O'Meara, B. C., Normark, B. B., Chung, J. H., & Jordal, B. H. (2001). The evolution of agriculture in beetles (Curculionidae: Scolytinae and Platypodinae). *Evolution*, **55**, 2011–2027.

Fraser, C. M., Gocayne, J. D., White, O. *et al.* (1995). The minimal gene complement of *Mycoplasma genitalium. Science*, **270**, 397–403.

Fritz-Laylin, L. K., Prochnik, S. E., Ginger, M. L. *et al.* (2010). The genome of *Naegleria gruberi* illuminates early eukaryotic versatility. *Cell*, **140**, 631–642.

Furuya, H. & Tsuneki, K. (2003). Biology of dicyemid mesozoans. *Zool. Sci.*, **20**, 519–532.

Gehring, W. J. (2005). New perspectives on eye development and the evolution of eyes and photoreceptors. *J. Heredity*, **96**, 171–184.

Graumann, P. L. (2007). Cytoskeletal elements in bacteria. *Ann. Review Microbiol.*, **61**, 589–618.

Harcet, M., Roller, M., Ćetković, H. *et al.* (2010). Demosponge EST sequencing reveals a complex genetic toolkit of the simplest metazoans. *Mol. Biol. Evol.*, **27**, 2747–2756.

Haring, E., Däubl, B., Pinsker, W., Kryukov, A., & Gamauf, A. (2012). Genetic divergences and intraspecific variation in corvids of the genus *Corvus* (Aves: Passeriformes: Corvidae) – a first survey based on museum specimens. *J. Zoological Systematics Evolutionary Research*, **50**, 230–246.

Hata, H. & Kato, M. (2006). A novel obligate cultivation mutualism between damselfish and *Polysiphonia* algae. *Biol. Lett.*, **2**, 593–596.

Heanue, T. A., Reshef, R., Davis, R. J. *et al.* (1999). Synergistic regulation of vertebrate muscle development by *Dach2*, *Eya2*, and *Six1*, homologs of genes required for *Drosophila* eye formation. *Genes Dev.*, **13**, 3231–3243.

Heino, J. (2007). The collagen family members as cell adhesion receptors. *BioEssays*, **29**, 1001–1010.

Hochberg, F. G. (1982). The "kidneys" of cephalopods: a unique habitat for parasites. *Malacologia*, **23**, 121–134.

Hofmann, M. A. (2001). Brain evolution in hominids: are we at the end of the road? In D. Falk & K. R. Gibson (eds.), *Evolutionary anatomy of the primate cerebral cortex*, pp. 113–127. Cambridge: Cambridge University Press.

Kant, S., Bagaria, D., & Ramakumar, S. (2002). Putative homeodomain proteins identified in prokaryotes based on pattern and sequence similarity. *Biochem. Biophys. Res. Comm.*, **299**, 229–232.

Keim, C. N. F., Abreu, F., Lins, U., Lins de Barros, H., & Farina, M. (2004). Cell organization and ultrastructure of a magnetotactic multicellular organism. *J. Struct. Biol.*, **145**, 254–262.

Kent, M. L., Andree, K. B., Bartholoew, J. L. *et al.* (2001). Recent advances in our knowledge of the Myxozoa. *J. Eukaryotic Microbiol.*, **48**, 395–413.

King, N., Westbrook, M. J., Young, S. L. *et al.* (2008). The genome of the choanoflagellate *Monosiga brevicollis* and the origin of metazoans. *Nature*, **451**, 783–788.

Kloepper, T. H., Kienle, C. N., & Fasshauer, D. (2007). An elaborate classification of SNARE proteins sheds light on the conservation of the eukaryotic endomembrane system. *Mol. Biol. Cell*, **18**, 3463–3471.

Kloepper, T. H., Kienle, C. N., & Fasshauer, D. (2008). SNAREing the basis of multicellularity: consequences of protein family expansion during evolution. *Mol. Biol. Evol.*, **25**, 2055–2068.

Kozmik, Z. (2008). The role of Pax genes in eye evolution. *Brain Research Bull.*, **75**, 335–339.

Kozmik, Z., Daube, M., Frei, E. *et al.* (2003). Role of Pax genes in eye evolution: a cnidarian PaxB gene uniting Pax2 and Pax6 functions. *Dev. Cell*, **5**, 773–785.

Kozmik, Z., Swamynathan, S. K., Ruzickova, J. *et al.* (2008). Cubozoan crystallins: evidence for convergent evolution of pax regulatory sequences. *Evol. Dev.*, **10**, 52–61.

Kuch, U., Müller, J., Mödden, C., & Mebs, D. (2006). Snake fangs from the Lower Miocene of Germany: evolutionary stability of perfect weapons. *Naturwissenschaften*, **98**, 84–87.

Laughlin, S. B., van Steveninck, R. R. de R., & Anderson, J. C. (1998). The metabolic cost of neural information. *Nature Neuroscience*, **1**, 36–41.

Li, Q., Chen, Y., & Yang, F. (2004). Identification of a collagen-like protein gene from white spot syndrome virus. *Archives Virology*, **149**, 215–223.

Lüttge, U. (2004). Ecophysiology of crassulacean acid metabolism (CAM). *Annals Botany*, **93**, 629–652.

McShea, D. W. (1998). Possible largest-scale trends in organismal evolution: eight 'live hypotheses'. *Annu. Rev. Ecol. Systematics*, **29**, 293–318.

McShea, D. W. (2002). A complexity drain on cells in the evolution of multicellularity. *Evolution*, **56**, 441–452.

Macklem, P. T. (2008). Emergent phenomenon and the secrets of life. *J. Applied Physiol.*, **104**, 1844–1846.

Marino, L., McShea, D. W., & Uhen, M. D. (2004). Origin and evolution of large brains in toothed whales. *Anatomical Record*, **281A**, 1247–1256.

Mineta, K., Nakazawa, M., Cebrià, F., Ikeo, K., Agata, K., & Gojobori, T. (2003). Original and evolutionary process of the CNS elucidated by comparative genomics analysis of planarian ESTs. *Proc. Natl. Acad. Sci. USA*, **100**, 7666–7671.

Mongodin, E. F., Nelson, K. E., Dougherty, S. *et al.* (2005). The genome of *Salinibacter ruber*: convergence and gene exchange among hyperhalophilic bacteria and archaea. *Proc. Natl. Acad. Sci. USA*, **102**, 18147–18152.

Nickel, M. (2010). Evolutionary emergence of synaptic nervous systems: what can we learn from the non-synaptic, nerveless Porifera? *Invertebrate Biol.*, **129**, 1–16.

Pecoits, E., Konhauser, K. O., Aubet, R. R. *et al.* (2012). Bilaterian burrows and grazing behavior at >585 million years ago. *Science*, **336**, 1693–1696.

Penrose, R. (1994). *Shadows of the Mind: a Search for the Missing Science of Consciousness*. Oxford: Oxford University Press.

Piraino, S., Zega, G., Di Benedetto, C. *et al.* (2011). Complex neural architecture in the diploblastic larva of *Clava multicornis* (Hydrozoa, Cnidaria). *J. Comparative Neurology*, **519**, 1931–1951.

Puigbò, P., Passamontes, A., & Garcia-Vallve, S. (2008). Gaining and losing the thermophilic adaptation in prokaryotes. *Trends Genetics*, **24**, 10–14.

Putnam, N. H., Srivastava, M., Hellsten, U. *et al.* (2007). Sea anemone genome reveals ancestral eumetazoan gene repertoire and genomic organization. *Science*, **317**, 86–94.

Royo, J. L., Maesto, I., Irimia, M. *et al.* (2011). Transphyletic conservation of developmental regulatory state in animal evolution. *Proc. Natl. Acad. Sci. USA*, **108**, 14186–14191.

Ryan, T. J. & Grant, S. G. N. (2009). The origin and evolution of synapses. *Nature Reviews. Neurosci.*, **10**, 701–712.

Sage, R. F. (2004). The evolution of C_4 photosynthesis. *New Phytol.*, **161**, 341–370.

Salthe, S. N. (2008). Natural selection in relation to complexity. *Artificial Life*, **14**, 363–374.

Schultz, T. R. & Brady, R. (2008). Major evolutionary transitions in ant agriculture. *Proc. Natl. Acad. Sci. USA*, **105**, 5435–5440.

Setiawan, J., Roccatagliata, V., Fedele, D. *et al.* (2012). Planetary companions around the metal-poor star HIP 11952. *Astronomy Astrophysics*, **540**, A141.

Srivastava, M., Simakov, O., Chapman, J. *et al.* (2010). The *Amphimedon queenslandica* genome and the evolution of animal complexity. *Nature*, **466**, 720–726.

Tcherkez, G. G. B., Farquhar, G. D., & Andrews, T. J. (2006). Despite slow catalysis and confused substrate specificity all ribulose bisphosphate carboxylases may be nearly perfectly optimized. *Proc. Natl. Acad. Sci. USA*, **103**, 7246–7251.

Thompson, D'Arcy, W. (1942). *On Growth and Form*. Cambridge: Cambridge University Press.

Valentine, J. W. (2000). Two genomic paths to the evolution of complexity in bodyplans. *Paleobiology*, **26**, 513–519.

Waller, L. N., Stump, N. J., Fox, K. F. *et al.* (2005). Identification of a second collagen-like glycoprotein produced by *Bacillus anthracis* and demonstration of associated spore-specific sugars. *J. Bacteriol.*, **187**, 4592–4597.

Wang, C.-S. & St. Leger, R. J. (2006). A collagenous protective coat enables *Metarhizium anisopliae* to evade insect immune responses. *Proc. Natl. Acad. Sci. USA*, **103**, 6647–6652.

Wickstead, B., Gull, K., & Richards, T. A. (2010). Patterns of kinesin evolution reveal a complex ancestral eukaryote with a multifunctional cytoskeleton. *BMC Evolutionary Biol.*, **10**, art. 110.

Zollikofer, C. P. E., Ponco de Léon, M. S., Lieberman, D. E. *et al.* (2005). Virtual cranial reconstruction of *Sahelanthropus tchadensis*. *Nature*, **434**, 755–759.

8 Evolution beyond Newton, Darwin, and entailing law: the origin of complexity in the evolving biosphere

Stuart A. Kauffman

My large aim in this chapter is to take us from our deeply received scientific world view and, derived from it, our view of the "real world" in which we live, that is, from the understanding of the world that was spawned by Newton and modern physics to an entirely different, newly vibrant, surprising, partially unknowable world of becoming in which the living, evolving world, biological, economic, and cultural co-creates, in an often unprestatable mystery, its own possibilities of becoming. If the latter perspective is right, we are beyond Newton, and even beyond Darwin, who, in all his brilliance, did not see that without natural selection "acting" at all to achieve it, the evolving biosphere creates its own future possibilities. And we will see, at the foundations of all this, that no laws entail the evolution of the biosphere, economy, or culture. But the biosphere is the most complex system we know in the universe. If it arose beyond entailing law, we must ask how this can be possible? More, is that "how" a hint to how complexity emerges at least in the living world, and perhaps in the abiotic universe?

We will begin to see ourselves in the living, evolving world in a world of inexplicable and unforeseeable opportunities that emerge with neither the "action" of natural selection in the evolving biosphere, or often without intent in the human world, that we partially co-create. It will follow that we live in not only a world of webs of

Complexity and the Arrow of Time, ed. Charles H. Lineweaver, Paul C. W. Davies and Michael Ruse. Published by Cambridge University Press.
© Cambridge University Press 2013.

cause and effect, but webs of opportunities that enable, but do not cause, often in unforeseeable ways, the possibilities of becoming of the biosphere, let alone human life. But, most importantly, I seek in this new world view a re-enchantment of humanity, of which this chapter may be a part. Our disenchantment following from Newton led to modernity. I believe we are partially lost in modernity, seeking half-articulated a pathway forward. Re-enchantment may be an essential part of this transformation.

I have many points to make and ideas to explore, and hope they will prove relevant and find resonance. If I am right, we are in the world in a way that we do not now clearly recognize.

8.1 NEWTON

How the Western, and the modern world, three hundred and fifty years later, changed with the inventions of one mind, Newton. He invented not only the mathematics, the differential and integral calculus, that gives us our way of thinking as moderns, from physics upward. He gave us his famous three laws of motion, and universal gravitation. Ask Newton: "I have six billiard balls rolling on a billiard table – what will happen to them?" and Newton might have rightly responded: "measure the positions and momenta and diameters of all the balls, the boundary conditions of the table, write down my three laws of motion in differential equation form representing the forces between the balls and between the balls and the edges of the table, then integrate my equations to yield the deterministic future trajectories of the balls". This ignores friction, of course, but classical physics handles that easily too.

What had Newton done? He had mathematized Aristotle's "efficient cause" in his differential equations giving forces between the entities, the laws of motion. He had invented a conceptual framework to derive the deterministic trajectory consequences by integration. But integration is deduction is "entailment", so the laws of motion in differential form entail the deterministic trajectories. In this entailment, Newton mathematized, in a very general framework,

Aristotle's argument that scientific explanation must be deduction. All men are mortal, Socrates is a man, hence, Socrates is mortal.

In the early 1800s, Simon Pierre Laplace further generalized Newton. Given a massive computing system, the Laplacian demon, informed of the instantaneous positions and momenta of all the particles in the universe, the entire future and (because Newton's laws are time-reversible) past of the universe is fully predictable and determined. This statement by Laplace is the birth of modern "reductionism", the long held view that there is some "final theory" down there, as in Stephen Weinberg's "Dream of a Final Theory", that will entail all that becomes in the universe.

We need two additional points. First, by the time of Poincaré, studying the orbits of three gravitating objects (a topic Newton knew was trouble), Poincaré was the first to show what is now known as deterministic chaos. Here tiny changes in initial conditions lead to trajectories which diverge from one another exponentially. Since we cannot measure positions and momenta to infinite accuracy, Poincaré showed that we cannot *predict* the behavior of a chaotic deterministic dynamical system. Determinism, contra Laplace, does not imply predictability. Second, quantum mechanics overthrew the ontological determinism of Newton, on most interpretations of quantum mechanics. Nevertheless, quantum systems obeying Schrödinger's equation *deterministically* evolve a *probability distribution* of the ontologically indeterminate probabilities of the actual, specific outcomes of quantum measurements. That evolution of the probability distribution is again entirely entailed.

With general relativity and quantum mechanics, the twin pillars of twentieth century physics were and remain firmly in place. No attempt to unite general relativity and quantum mechanics has been successful after eighty-five years of trying. Further, no attempt to deduce the specific outcome of a quantum measurement from within quantum mechanics has succeeded either. Success may or may not come. In modern physics, the conviction remains that all that arises in the universe is entailed. I note, however, that if the specific outcome of a quantum measurement cannot be deduced from within quantum mechanics, that specific outcome, hence the becoming of the universe via all those specific measurement outcomes and their consequences, is NOT entailed. (I thank R. Melmon for useful recent discussion.) If so, no final theory entails all that becomes in the universe.

8.2 DARWIN

After Newton, and perhaps as profoundly, Darwin changed our thinking. We all know the central tenets of his theory: heritable variation among a population, competition for resources insufficient for all to survive, and hence, natural selection favoring those variants "fitter" in the current environment. Thus, we achieve "adaptation", and critically, the *appearance of design* without a designer. The well-known story of the difficulties of Darwin's theory with "blending inheritance" and its unexpected rescue by Mendelian genetics, even the fact that a copy of Mendel's work lay unopened on Darwin's desk, is well known. Mendelian genetics prevents blending inheritance and paved the way for the mid-twentieth century neo-Darwinian – or "modern" – synthesis.

The entire panoply of life's evolution at last lay open to at least the start of understanding, given Darwin. The history from Darwin and Mendel to the neo-Darwinian synthesis of the mid-twentieth century is well known. I would comment briefly here that the neo-Darwinian synthesis left out Waddington's "epigenesis" – the attempt, starting with Wolfe, to link development and evolution. Just as the mathematical inventions of population genetics by Fisher, Haldane, and Wright served to unite Mendel and Darwin, the inventions of mathematical models of the genetic regulatory networks envisioned by Waddington have begun to fuse the neo-Darwinian synthesis with developmental biology. For example, my own early study of random Boolean networks as models of genetic regulatory networks, (Kauffman, 1969, 1993), may have helped start this movement. And now systems biology, including the modern sense of epigenetics as heritable changes in chromosomal "markings" by methylation or acetylation of histones, without changing the DNA sequence, are further parts of extending the neo-Darwinian synthesis to include development, environment, and ecology, into a broadened synthesis.

8.3 MONOD AND "TELEONOMY"

The concept of "function", "doing", "purpose", and "agency" in biology and, with it, a potential "meaning" for signs or symbols, totally absent in physics where only "happenings" occur, has been mooted in standard biology by a concept voiced by Jaques Monod (Monod, 1995). Consider a bacterium swimming up a glucose gradient. It "seems" to be "acting to get food". But, said Monod, this view of the organism seems to be entirely wrongheaded. The cell in its environment is just an evolved molecular machine. Thanks to natural selection, the swimming up the gradient gives the appearance of purpose, of teleology, but this is false. Instead, this behavior is a mere "as if" teleology that Monod called "teleonomy". (In a wider view, Monod struggled to reconcile teleonomy with his sense that organisms do act with purpose, which issue he thought was the central issue of biology.) In short, for Monod, via teleonomy, and for legions of later biologists and philosophers, "doing" is unreal in the universe, only the mechanical, selected appearance of "doing" is real.

Indeed, in so arguing, Monod is entirely consistent with physics. As noted, there are no "functions", "doings", "meanings", or "agency" in physics. Balls rolling down a hill are merely Newtonian "happenings". So, too, are the happenings in the evolved molecular machine that is the bacterium swimming up the glucose gradient. Yet we humans think functions and doings and agency are real in our world. If so, from whence came functions, doings, agency, and meanings? Are, indeed, functions, meanings, and doings real in the universe? I now give, as far as I know, an entirely new set of arguments that, I believe, fully legitimizes functions, doings, agency, and even meanings as real in the universe, but beyond physics. The discussion has a number of steps.

8.4 BEYOND TELEONOMY: FUNCTION, DOING, AGENCY

8.4.1 The non-ergodic universe above the complexity of the atom

Has the universe in 13.8 billion years of existence created all the possible fundamental particles and stable atoms? Yes. Now consider proteins. These are linear sequences of twenty kinds of amino acids that typically fold into some shape and catalyze a reaction or perform some structural or other function. A biological protein can range from perhaps fifty amino acids long to several thousands. A typical length is three hundred amino acids long. Then let us consider all possible proteins that are two hundred amino acids in length. How many are possible? Each position in the two hundred has twenty possible choices of amino acids, so there are $20 \times 20 \times 20 \ldots 200$ times, or 20 to the 200th power, which is roughly 10 to the 260th power possible proteins of 200 amino acids in length:

Now let us ask if the universe can have created all these proteins since its inception 13.8 billion years ago. There are roughly 10 to the 80th particles in the known universe. If these were doing nothing, ignoring space-like separation, but making proteins on the shortest timescale in the universe, the Planck timescale of 10 raised to the −43 seconds, it would take 10 raised to the 39th power times the lifetime of our universe to make all possible proteins of length 200, just *once*. In short, in the lifetime of our universe, only a tiny fraction of all possible proteins of length 200 can have been created. This means profound things. First, the universe is vastly non-ergodic in the physicists' sense of the ergodic hypothesis at the foundation of statistical mechanics. It is not like a gas at equilibrium in statistical mechanics. With this vast non-ergodicity, when the possibilities are vastly larger than what can actually happen, history enters. Not only will we not make all possible proteins of length 200 or 2000, we will not make all possible organs, organisms, social systems,... There is an *indefinite* hierarchy of non-ergodicity as the complexity of the objects we consider increases.

I note that Conway Morris in Chapter 7 considers convergent evolution, which happens often, and concludes that the space of phenotypes is "saturable" and has been quite saturated by evolution. Were the "tape" played again, we would get much the same organisms. I agree that convergent evolution is powerful, but given the non-ergodicity of the universe at the level of complex molecules, organs, and organisms, where the non-ergodicity of the universe increases as the complexity of the "objects" (molecules, protocells, cells, organisms) increases, I strongly doubt Conway Morris's interesting "saturation" claim. This remains an important open question.

8.4.2 Kantian wholes and the reality of functions, "doings, and agency"

The great philosopher, Immanuel Kant, wrote (Kant, 2000) that "in an organized being, the parts exist for and by means of the whole, and the whole exists for and by means of the parts." Kant was at least considering organisms, which I will call Kantian wholes. Functions are clearly definable in a Kantian whole. The function of a part is its causal role in sustaining the existence of the Kantian whole. Other causal consequences are side effects. Note that this definition of function rests powerfully on the fact that Kantian wholes, like a bacterial cell dividing, are complex entities that *only get to exist in the non-ergodic universe above the level of atoms because they are Kantian self-recreating, non-equilibrium, wholes*. It is this combination of self-recreation of a Kantian whole, and therefore its very existence in the non-ergodic universe above the level of atoms that, I claim, fully legitimizes the word, "function" of a part of a whole in an organism. Functions are real in the universe.

Now consider the bacterium swimming up the glucose gradient to "get food", Monod's merely teleonomic *as if* "doing". But we can rightly define a behavior that sustains a Kantian whole, say the bacterium existing in the non-ergodic universe, as a "doing". Thus, I claim, "doings" are real in the universe, not merely Monod's teleonomy. Furthermore, "agency" enters with "doing". In my third book,

Investigations (Kauffman, 2000), my own attempt to define agency stated that a molecular autonomous agent was a self-reproducing system that is able to do at least one thermodynamic work cycle. With Philip Clayton, we broadened this definition to involve the inclusion of the self-reproducing system in some boundary, say a liposome, and the capacity to make at least one discrimination, food or not food, and to "act" upon that discrimination (Kauffman & Clayton, 2006). Bacteria clearly do this, and, without invoking consciousness, are therefore agents. Agency is real in the universe.

A rudimentary beginning of "emotion" emerges here (Piel, 2012). The bacterium must sense its world and act to avoid toxins and to obtain food. The evaluation of "good" versus "bad", arguably the "first sense", enters here. Agency and the existence of the cell precedes this "semiotic" evaluation logically, for if there were no existence in the non-ergodic universe of the Kantian whole, agency and evaluation of food versus poison would not be selected, and so they would not exist in the universe. Thus, I would argue that life is not sufficiently based on semiosis, for, as noted, if there were not a prior Kantian whole existing in the non-ergodic universe above the level of atoms, semiosis would not have evolved. I note also that Hume's famous: "one cannot deduce an 'ought' from an 'is'", the famous naturalistic fallacy, rests on the critical fallacy that Hume, like Descartes, thought of a mind "knowing" its world. Hume did not think of an agent "acting" in its world. Given "action" and "doing", doing "it" well or poorly enters inevitably, and with it, "ought". With ought, the need for evaluation, the rudiments of emotion without positing consciousness, enter.

Interestingly, Kant opined that there would never be a Newton of biology. Despite Darwin, a major point of this chapter, which will take us beyond physics, is that here Kant was right. There never, indeed, will be a Newton of biology, for, as we will see below, unlike physics and its law-entailed trajectories, the evolution of the biosphere cannot be entailed by laws of motion and their integration.

8.4.3 Collectively autocatalytic DNA sets, RNA sets, or peptide sets

Gonen Ashkenasy (Askenasy & Wagner, 2009) at the Ben Gurion University in Israel has created in the laboratory a set of nine small proteins, called peptides. Each peptide speeds up, or catalyzes, the formation of the *next* peptide by ligating two fragments of that next peptide into a second copy of itself. This catalysis proceeds around a *cycle* of the nine peptides. It is essential that in Ashkenasy's real system, no peptide catalyzes its *own* formation. Rather the set as a whole *collectively catalyzes its own formation*. I shall call this a collectively autocatalytic set, "CAS".

These astonishing results prove a number of critical things. First, since the discovery of the famous double helix of DNA, and its Watson–Crick template replication, many workers have been convinced that molecular reproduction *must* rest on something like a template replication of DNA, RNA, or related molecules. It happens to be true that all attempts to achieve such replication without an enzyme have failed for 50 years. Ashkenasy's results demonstrate that small proteins can collectively reproduce. Peptides and proteins have no axis of symmetry like the DNA double helix. These results suggest that molecular reproduction may be far easier than we have thought. I mentioned that in Kauffman (1971, 1986, 1993). I invented a theory for the statistically expected emergence of collectively autocatalytic sets in sufficiently diverse "chemical soups". This hypothesis, tested numerically, is now a theorem (Mossel & Steel, 2005). If so, routes to molecular reproduction in the universe may be abundant. RNA collectively autocatalytic sets (Lam & Joyce, 2009) and DNA collectively autocatalytic sets (von Kiedrowski, 1986) have been created experimentally.

I raise a new question. We must ask: What kind of "law" does this theory of the spontaneous emergence of collectively autocatalytic sets involve? We will see that whatever form this "law" may be, it is *not* a Newton-like law with initial and boundary conditions, laws of motion in differential equation form, and the integration of those laws yielding entailed trajectories – i.e., our hallmark case since Newton of "mechanism". Newton mathematized "efficient cause", one of Aristotle's four causes: material, final, formal, and efficient. I suspect that the theory of the spontaneous emergence of collectively autocatalytic sets is a new kind of law: a formal cause law. If this is right, one, and perhaps the best, theory for the emergence of life as an expected property in ensembles of chemical reaction networks is *not* a mechanistic law at all, and the emergence of such sets is beyond mechanism in a new sense, even though each instantiation of such an emergence in the *ensemble* of all possible chemical reaction networks has a set of mechanisms. Collectively autocatalytic DNA sets and RNA sets have also been made.

8.4.4 Collectively autocatalytic sets are the simplest cases of Kantian wholes and the peptide parts have functions

A collectively autocatalytic set is precisely a Kantian whole, which "gets to exist" in the non-ergodic universe above the level of atoms, precisely because it is a self-reproducing, non-equilibrium, Kantian whole. Moreover, given that whole, the "function" of a given peptide part of the nine peptide set is exactly its role in catalyzing the ligation of two fragments of the next peptide into a second copy of that peptide. The fact that the first peptide may jiggle water in catalyzing this reaction is a causal side effect that is *not* the function of the peptide. Thus, functions are typically a subset of the causal consequences of a part of a Kantian whole.

8.4.5 Task closure

Collectively autocatalytic sets exhibit a terribly important property. If we consider catalyzing a reaction a "catalytic task", then the set as a *whole* achieves "task closure". All the reactions that must be catalyzed by at least one of Ashkenasy's nine peptides *are* catalyzed by at least one of those peptides. No peptide catalyzes its own formation. The set as a whole catalyzes its own reproduction via a clear task closure.

8.4.6 Task closure in a dividing bacterium

Consider a dividing bacterium. It, too, achieves some only partially known form of task closure in part in, and via, its environmental niche. But the tasks are far wider than mere catalysis. Among these tasks are DNA replication, membrane formation, the formation of chemosmotic pumps, and complex cell signaling mechanisms in which a chemically *arbitrary* molecule in the environment can bind to part of a trans-membrane protein, and thereby alter the behavior of the intracellular part of that molecule, which, in turn, unleashes intracellular signaling. Thus this task closure is over a wide set of tasks.

8.4.7 Biosemiosis enters at this point

I thank Professor Kalevli Kull of the Tartu University Department of Semiotics for convincing me that, at just this point, biosemiotics enters. As Kull points out, the set of environmental molecules that can bind the outside parts of transmembrane proteins are chemically arbitrary – a point Monod emphasized as well in considering allosteric enzymes. Thus, as Kull further points out, the set of states of the different molecules outside the cell that can bind to the outside parts of these transmembrane proteins and unleash intracellular signaling and a coordinated cellular response constitute a *semiotic code* by which the cell navigates its "known" world, "known" – without positing consciousness – via the code and, in general, probably evolved by selection encoding of the world as "seen" by the organism. Change the molecule species binding the outside of the transmembrane proteins, and the world the cell "knows" and evaluates changes. Biosemiosis is real in the universe.

8.5 TOWARD: NO ENTAILING LAWS, BUT ENABLEMENT IN THE EVOLUTION OF THE BIOSPHERE

I now shift attention to a new, and I believe, transformative topic. With my colleagues Giuseppe Longo and Mael Montevil, mathematicians at the École Polytechnique in Paris, I wish to argue that *no law* entails the evolution of the biosphere (Longo, Montevil, & Kauffman,

2012). If we are right, entailing law, the centerpiece of physics since Newton, ends at the watershed of evolving life. If this claim is right, it is obviously deeply important. Furthermore, it raises the issue of how the biosphere, the most complex system we know in the universe, can have arisen beyond entailing law. I will discuss these issues as well. And, central to this book, *if* entailing law is the instantiation of Descartes' *Res extensa* and mechanistic worldview as mathematized by Newton's "entailing" laws of motion, we will find a new sense in which evolution is not mechanism. Again, the foregoing discussion proceeds in several steps.

8.5.1 *The uses of a screwdriver cannot be listed algorithmically*

Here is the first "strange" step. Can you name all the uses of a screwdriver, alone, or with other objects or processes? Well, screw in a screw, open a paint can, wedge open a door, wedge closed a door, scrape putty off a window, stab an assailant, be an objet d'art. The screwdriver tied to a stick can be a fish spear, the spear rented to "natives" for a 5% fish catch return becomes a new business,...I think we all are convinced that the following two statements are true: (1) the *number* of uses of a screwdriver is *indefinite*; and (2) unlike the integers which can be ordered, there is no natural ordering of the uses of a screwdriver. The uses are *unordered*. But these two claims entail that there is no "Turing effective procedure" to *list* all the uses of a screwdriver alone or with other objects or processes. In short, there is no algorithm to list all the uses of a screwdriver.

Now consider *one* use of the screwdriver, say to open a can of paint. Can you list all the other objects, alone or with other objects or processes that may carry out the "function" of opening a can of paint? Again, the number of ways to achieve this function are indefinite in number, and unorderable, so again, no algorithm can list them all.

8.5.2 *Adaptations in an evolving cell cannot be prestated*

Now consider an evolving bacterium or eukaryotic, say, single celled organism. In order to adapt in some new environment, all that has to occur is that some one or many cellular or molecular "screwdrivers" happen to "find a use" that enhances the fitness of the evolving cell in that new environment. Then there must be heritable variation for those properties of the cellular screwdrivers, then natural selection, acting at the level of the Kantian whole cell, not at the level of the celllular screwdriver itself, will select the fitter variant cells with the new uses of the molecular screwdrivers which constitute adaptation.

I wish to make an additional point here. It is widely known that Darwin gave us natural selection "culling", and something like the "survival of the fittest", the appearance of design without a designer. But from whence came "the *arrival* of the *fittest*", or at least the *fitter*? The above selection at the level of the Kantian whole cell of cellular or molecular screwdrivers for a new or modified use *is* the arrival of the "fitter". This seems deeply important, for evolution has lacked an account of "the arrival of the fittest, or fitter", that is, whence comes adaptation? Adaptation consists in just this "finding of a use" that enhances fitness. More, by the above discussion, we cannot *prestate* what this use will be. No algorithmic *list* of the possible uses of these cellular and molecular screwdrivers can be had; thus we cannot know, ahead of time, what natural selection *acting at the level of the Kantian whole organism*, will reveal as the new uses of the cellular screwdrivers acting in part via the niche of the organism, which *succeed better*, hence were selected. We cannot, in general, prestate the adaptive changes that will occur. This is the deep reason evolutionary theory is so weakly predictive.

8.5.3 We cannot prestate the actual niche of an evolving organism

The task closure of the evolving cell is achieved, in part, via causal or quantum consequences passing through the environment that constitute the "actual niche" of the evolving organism. But the features of the environmental "niche" that participate with the molecular screwdrivers in the evolving cell which will allow a successful task closure are circularly defined with respect to the organism itself. We only know after the fact of natural selection what aspects of the evolving cell, its screwdrivers, as well as which of the causal consequences of specific aspects of the actual niche, are successful when selection has acted at the level of the Kantian whole evolving cell population. Thus, we cannot prestate the actual niche of an evolving cell by which it achieves task closure in part via that niche.

But these facts have deep meaning. In physics, the phase space of the system is *fixed*, in Newton, Einstein, and Schrödinger. This allows for entailing laws. In evolution, each time an adaptation occurs and a molecular or other screwdriver finds a new use in a new actual niche, the very phase space of evolution has *changed* in an unprestatable way. But this means that we can *write no equations of motion for the evolving biosphere*. More, the actual niche can be considered as the boundary conditions on selection. But we cannot prestate the actual niche. In the case of the billiard balls, Newton gave us the laws of motion, and told us to establish initial and boundary conditions and then integrate laws of motion stated in differential equation form to get the entailed trajectories. But, in biology, we cannot write down the laws of motion, so we cannot write them down in differential equation form. Even if we could, we could not know the niche boundary conditions, so we could not integrate those laws of motion that we do not have anyway. It would be like trying to solve the billiard ball problem on a billiard table whose shape changed forever in unknown ways. We would thereby have no mathematical model. Here, the profound implication is that *no laws entail the evolution of the biosphere*.

If this is correct, we are, as stated above, at the end of reductionism at the watershed of evolving life. Now the machine metaphor since Descartes, perfected by Newton, leads us to think of organisms, as Monod stated, as molecular *machines*. Let me distinguish diachronic from synchronic science. Diachronic science studies the evolution of life, and its "becoming", over time. Synchronic science studies the presumably fully reducible aspects of, for example, how a heart, once it has come to exist in the non-ergodic universe, "works". But in the diachronic becoming of the *biosphere, life is an ongoing, unprestatable, non-algorithmic, non-machine, non-equilibrium* process creating "fitter" variants (the unprestatable arrival of the fittest), then selecting at the level of the Kantian whole for that increased fitness variant.

There are, I think, deep implications to this. What is the "phenotype"? This is a much debated question. But few would doubt that the causal consequences of a part that sustains the Kantian whole are its *function*, and that, therefore, *functions are either the phenotype or at least part of the phenotype*. This realization has an odd and new consequence. When some cellular or molecular screwdrivers alone or together "find some use" in the current, or new, environment that enhances the fitness of the Kantian whole organism, and so are selected if heritably variable, in a deep sense it is "immaterial" just what cellular or molecular screwdrivers, and which of their specific causal consequences, happen to fulfill the new adaptive use. Thus, in a deep sense, the arrival of the fitter does not depend in any specific way on any specific predefinable set of cellular or molecular screwdrivers. Although any specific instantiation of an adaptive step to a fitter function or new function will in fact utilize some specific cellular or molecular screwdrivers, which ones may happen to be the ones so selected cannot be algorithmically prestated. Thus, not only is there the familiar "multiple realizability" philosophers are used to, the situation is more radical – the multiple realizations for a given new use cannot be algorithmically listed, so are unprestatable. In this sense, the adaptive change of the organism is beyond entailing law and beyond stable mechanism.

8.6 DARWINIAN PREADAPTATIONS AND RADICAL EMERGENCE: THE EVOLVING BIOPSHERE, WITHOUT THE "ACTION" OF SELECTION, CREATES ITS OWN FUTURE POSSIBILITIES OF BECOMING

If we asked Darwin what the function of my heart is, he would respond, "to pump your blood". But my heart makes heart sounds and jiggles water in my pericardial sac. If I asked Darwin why these are not the function of my heart, he would answer that I have a heart because its pumping of blood was of selective advantage in my ancestors. In short, he would give a selection account of the causal consequence for virtue of which I have a heart. Note that he is also giving an account of why hearts exist at all as complex, non-equilibirum, entities in the non-ergodic universe above the complexity of atoms. Hearts are functioning parts by pumping blood, of humans as reproducing Kantian wholes. Note again that the function of my heart is a subset of its causal consequences, pumping blood, not heart sounds or jiggling water in my pericardial sac.

Darwin had an additional deep idea: a causal consequence of a part of an organism of no selective significance in a given environment might come to be of selective significance in a different environment, so be selected, and, typically, a new function would arise. These are called "Darwinian preadaptations" without meaning foresight on the part of evolution. Stephen Jay Gould renamed them "exaptations". I give but two examples of thousands of Darwinian preadaptation. First, your middle ear bones that transmit sound from the ear drum to the oval window and cochlea of the inner ear evolved by Darwinian preadaptations from the jaw bones of an early fish. Second, some fish have a swim bladder, a sac partly filled with air and partly with water, whose ratio determines neutral buoyancy in the water column. Paleontologists believe that the swim bladder evolved from the lungs of lung fish. Water got into some lungs, now sacs partly filled with air, and partly with water, were poised to evolve into swim bladders. Let us assume the paleontologists are right.

I now ask three questions: firstly, did a new function come to exist in the biosphere? Yes, hearing and neutral buoyancy in the water column. Secondly, did the evolution of the middle ear or swim bladder alter the future evolution of the biosphere? Yes, new species of animals with hearing and new species of fish with swim bladders evolved. They evolved new mutant proteins. And, critically, the middle ear, or the swim bladder, *once each came to exist*, constituted what I will call *a new adjacent possible empty niche*, for a worm, bacterium or both could evolve to live only in the middle ear or swim bladders. Thus the middle ear or swim bladder, once each exists, alters the possible future evolution of the biosphere. I return to this point in a moment for enchantment hides here. Third, now that you are an expert on Darwinian preadaptations, can you name all possible Darwinian preadaptations just for humans in the next three million years? Try it and feel your mind go blank. We all say "No". A start to why we cannot is to ask the following questions: how would you name all possible selective environments? How would you know you listed them all? How would you list all the features of one or many organisms that might serve as "the preadaptation"? We cannot.

The underlying reasons why we cannot do this is given above in the discussion about screwdrivers. As we saw, we cannot algorithmically list all their uses, either alone or with other objects and processes. In addition, as we saw, we cannot algorithmically list the other objects and processes that can accomplish any specific task, e.g. opening a can of paint, that we can use a screwdriver to accomplish. In addition, the organism completing task closure in part via its actual niche is circularly defined and cannot be prestated until selection at the level of the Kantian whole organism reveals what causal aspects of what screwdriver parts have "worked", hence what the relevant variables, functions, and aspects of the environment actually now are. These are the deep reasons we cannot list all possible selective environments, and all possible preadaptations. Because we cannot "effectively list" these, we cannot know what the set of all the possibilities of the evolution of the biosphere are. Not only do we not know what WILL happen, we don't even know what CAN happen.

8.6.1 The adjacent possible

Consider a flask of one thousand kinds of small organic molecules. Call these the actual. Now let these react by a single reaction step. Perhaps new molecular species may be formed. Call these new species the molecular "adjacent possible". It is perfectly defined if we specify a minimal stable lifetime of a molecular species. Now, let me point at the adjacent possible of the evolving biosphere. Once lung fish existed, swim bladders were in the adjacent possible of the evolution of the biosphere. But two billion years ago, before there were multicelled organisms, swim bladders were not in the adjacent possible of the evolution of the biosphere.

I think we all agree to this. But now consider what we seem to have agreed to: with respect to the evolution of the biosphere by Darwinian preadaptions, *we do not know all the possibilities*. Now let me contrast our case for evolution with that of flipping a fair coin 10000 times. Can we calculate the probability of getting 5640 heads? Sure, use the binomial theorem. But note that here we know *ahead of time* all the possible outcomes, all heads, all tails, alternative heads and tails, all the 2 to the 10000 power possible patterns of heads and tails. Given that we know all the possible outcomes, we thereby know the "sample space" of this process, so can construct a probability measure. We do not know what *will* happen, but we know what *can* happen.

But in the case of the evolving biosphere, not only do we not know what *will* happen, we do not even know what *can* happen. There are at least two huge implications of this: firstly, we can construct no probability measure for this evolution by any known mathematical means. We do not know the sample space. Secondly, reason, the prime human virtue of our Enlightenment, cannot help us in the case of the evolving biosphere, for we do not even know what *can* happen, so we cannot reason about it. The same is true of the evolving econo-sphere, culture, and history; as I will try to show us, we often do not know ahead of time the new variables which will become relevant, so we cannot reason about them. Thus, real life is not an optimization problem, top down, over a known space of possibilities. It is far more mysterious. How do we navigate, not knowing what can happen? Yet we do. This has very large implications for how we govern ourselves and live wisely when we cannot know all that *can* happen.

8.6.2 Without natural selection, the biosphere enables and creates its own future possibilities

Now I introduce a radical emergence that I find re-enchanting. Consider the middle ear, or swim bladder, once either has evolved. We agreed above, I believe, that a bacterium or worm or both could evolve to live only in that middle ear or swim bladder, so the middle ear or swim bladder as *a new adjacent possible empty niche*, once it had evolved, alters the future possible evolution of the biosphere. Next, did natural selection act on an evolving population of hearing animals or fish to select a well-functioning middle ear system or swim bladder? Of course (I know I am here anthropomorphizing selection, but we all understand what is meant). But did natural selection "act" to create the middle ear or swim bladder as *a new adjacent possible empty niche*? No! Selection did not "struggle" to create the middle ear or swim bladder as a new empty adjacent possible niche. But that means something I find stunning. Without selection acting in any way to do so, evolution is creating its own future possibilities of becoming! And the worm or bacterium or both that evolves to live in the middle ear or swim bladder is a radical emergence unlike anything in physics.

It seems important to stress that the new realization that the biosphere, without natural selection "acting" to achieve it, creates its own future possibilities of becoming, hence radical emergence into those ever new and unprestatable "adjacent possible" empty niches, was not seen by Darwin, even with his insights into preadaptations, nor by contemporary evolutionary theory, including the neo-Darwinian synthesis. We are, with "no laws entail the evolution of the biosphere", if true, beyond Newton, Einstein, and Schrödinger at the watershed of evolving life. And with the enchantment of the fact that the evolving biosphere creates, beyond selection, its own future possibilities, we are beyond Darwin. We have entered an entirely new worldview. It is a worldview of unprestatable, beyond entailing law becoming, where evolution, without selection doing so, creates its own future, unprestatable possibilities of becoming. Note that I am *not* talking here about familiar "niche construction", where the behaviors of organisms modify their niches, nor am I talking about the Baldwin effect. I am talking about typically unprestatable "niche creation" itself.

8.7 EVOLUTION OFTEN DOES NOT *CAUSE*, BUT *ENABLES* ITS FUTURE EVOLUTION

The bacterium or worm that evolves to live in the actual niche of the middle ear or swim bladder whereby it achieves a task closure selected at the level of the Kantian whole worm or bacterium, evolves by quantum indeterminate and ontologically acausal quantum events, later selected, if heritable and fitter, by natural selection acting at the level of the whole organism in its world. Thus, the swim bladder does not *cause*, but *enables, that is, "makes possible"*, the evolution of the bacterium or worm or both to live in the swim bladder. This means that evolving life is not only a web of cause and effect, but of empty niche opportunities, that enable new evolutionary radical emergence as the evolving biosphere creates, beyond selection, its own future possibilities. The same is true in the evolving econosphere, cultural life, and history, as I discuss more fully below. We live in both a web of cause and effect and a web of enabling opportunities that enable new possible directions of becoming.

More, the swim bladder or middle ear as an "adjacent possible" empty niche opportunity for adaptation by acausal quantum mutation events then selected, if heritable, is itself, as that new adjacent possible niche, an "enabling constraint". As that enabling constraint, the new niche shapes evolution and enables the worm or bacterium in the adaptive solution it finds in the new adjacent possible niche created by that new niche as an "enabling constraint". Thus evolution is not just a web of cause and effect, but also a web of niche opportunities, arising without selection "acting" to achieve them, that enable new adjacent possibilities for the becoming of the biosphere.

The same claim is true for the entire evolution of the multi-specied biosphere. Each species or set of species plus abiotic environments creates enabling constraint opportunities, new adjacent possible niches, for the evolution of yet new species. This raises a fascinating new question: can we find an account of how enabling constraints generate new adjacent possible ways of life and shape them, thereby generating yet new enabling constraints that enable new adjacent possible empty niches for yet new species as evolution unfolds?

Perhaps, in some hard to define sense, evolution "maximizes" the average growth of its own adjacent possibilities into which it "becomes".

8.7.1 Neither quantum mechanics alone nor classical physics alone account for evolution

Mutations are often quantum acausal and indeterminate, random events. Yet evolution is not random: the eye evolved some eleven times and the vertebrate and octopus camera eyes are nearly identical (except that the blood vessels in the octopus are behind the retina). These examples, widespread, of convergent evolution show that evolution is not random. Thus neither quantum mechanics alone nor classical physics alone suffices to account for evolution.

8.8 TOWARD A POSITIVE SCIENCE FOR THE EVOLVING BIOSPHERE BEYOND ENTAILING LAW

The arguments above support the radical claim that no laws entail the evolution of the biosphere. If right, Kant was right. There will be no Newton of biology. Not even Darwin was that Newton yielding entailing laws. But the biosphere is the most complex system we know in the universe, and it has grown in diversity and flourished, even with small and large avalanches of extinction events, for 3.8 billion years. Indeed, there was a secular increase in species diversity over the Phanerozoic.

How are we to think of the biosphere building itself, probably beyond entailing laws? Organisms are Kantian wholes, and the building of the biosphere of these past 3.8 billion years seems almost certainly to be related to how Kantian wholes co-create their worlds with one another, including the creating, with no natural selection "acting" to so achieve it, of new empty adjacent possible niches that alter the future possible evolution of the biosphere. There may be a way to start studying this topic: a new quest. Collectively, autocatalytic sets are the simplest models of Kantian wholes. In very recent work with Wim Hordijk and Michael Steel, computer scientist and mathematician, respectively (Hordijk, Steel, & Kauffman, 2012), we are studying what they call RAFs, which are defined as collectively autocatalytic sets in which the chemical reactions, without catalysis, occur spontaneously at some slow finite rate for all reactions, including reactions from the "food set" of exogenously maintained "food molecules" that are inputs to the set, but that reaction rate is much speeded up by catalysis. Fine results by Horkijk and Steel show that RAFs emerge and require only that each catalyst catalyzes between one and two reactions. This is fully reasonable both chemically and biologically (Hordijk & Steel, 2004).

Most recently, the three of us have examined the substructure of RAFs (Hordijk, Kauffman, & Steel, 2012). There are irreducible RAFs, which, given a food set of sustained small molecules, have the property of autocatalysis, but if any molecule is removed from the RAF the total system collapses. It is irreducible. Then, given a maximum length of polymers allowed in the model as the chemicals, from monomers to longer polymers, there is a maximal RAF, which increases as the length of the longest allowed polymer, and hence the total diversity of possible polymers allowed, increases. The most critical issue is this: there are *intermediate* RAFs called "submaximal RAFs", each composed either of two or more irreducible RAFs or of one or more irreducible RAF and one or more larger "submaximal" RAF, or composed of two or more smaller submaximal RAFs. Thus we can think mathematically of the complete set of irreducible RAFs, all the diverse submaximal RAFs, and the maximal RAF. For each, we can draw arrows from those smaller RAFs that jointly comprise it. This set of arrows is a partial ordering among all the diverse RAFs possible in the system.

The next important issue is this: if new food molecule species, or larger species, enter the environment, even *transiently*, the total system can grow to create NEW submaximal RAFs that did not exist in the system before. This is critical. It shows that existing Kantian wholes can create new empty adjacent possible niches and with a chemical fluctuation in which molecular species are transiently present in the environment, the total "ecosystem" can grow in diversity. A model biosphere is building itself! Thus simpler collectively autocatalytic sets can evolve into more complex ones.

In this system, the diverse RAFs can "help" one another, for example, a waste molecule of one can be a food molecule of another, or via inhibition of catalysis, or toxic products of one with respect to another, can hinder one another in complex ways. They form a complex ecology. Further, these RAFs, if housed in compartments that can divide, such as bilipid membane vesicles called liposomes (Luisi, Mavelli, Rasi, & Stano, 2004), have been shown recently to be capable of open ended evolution via natural selection, where each of the diverse RAFs acts as a "replicator" to be selected, and in that selection, chemical reaction "arcs" that flower from the RAF core act as the phenotype with the core. Thus, to my delight, we have the start of a theory for the evolution of Kantian wholes into ecosystems of now interacting Kantian wholes which may, in fact, require one another for their joint existence. In this case, we must redefine "whole".

But there is a profound limitation to these models: they are in a deep sense algorithmic and their possible phase spaces can be prestated. The reason is simple: the only functions that happen in these RAF systems are molecules undergoing reactions, which are catalyzed by molecules. But the set of possible molecules up to any maximum length polymer can be prestated. And the set of possible catalytic interactions can be prestated, therefore, even in models where the actual assignment of which molecule catalyzes which reaction is made at random or via some "match rule" of catalyst and substrate(s), then assigned with a probability of catalysis per molecule, thus we have already prestated all the possibilities for reaction systems and all possible patterns of catalysis. Therefore, the phase space is prestated. This may "hint" at new principles, as I try to say below, for the burgeoning complexity of the universe.

By contrast, in the discussion above, we talked about the vast task closure achieved by an evolving bacterium or eukaryotic cell or organism. These tasks were not limited to catalysis. And, as we saw with the discussion of the possible uses of a molecular screwdriver in a cell, those uses are both indefinite in number and not orderable, so no algorithm can list all those uses. Nor can we prestate how the evolving Kantian whole cell will evolve via natural selection, where selection acts at the level of the Kantian whole and favors whole cells whose altered screwdriver parts with heritable variations achieve some often new, fitter, functional task closure via the actual niche. Thus the real evolutionary process is non-algorithmic, non-machine, non-entailed. With respect to our initial evolving RAF ecosystems, we do not yet know how to make this evolution non-algorithmic and non-entailed. While we have a start, and a useful one, it is not enough.

8.8.1 The economy is an evolving autocatalytic set

In the theory of autocatalytic sets, typically modeled are the "molecules" as binary strings (1000100) as substrates, products, and catalysts for speeding the reactions among the substrates and products. BUT there is nothing at all that is fundamentally "chemical" about the binary symbol strings (e.g., 10001010). These symbol strings can stand for any objects that can "react" and transform. Thus consider the physical transformations in the real economy. Here two boards and a nail act as "substrate inputs" to a "production capacity" consisting of a hammer and a human hand. The production capacity is the analog of the reaction among chemical substrates to produce chemical products. Here, the two boards and nail as input goods are "acted upon, catalytically" by the hammer and hand to produce the product: two boards nailed together. Now note that the hammer itself is a product of the economic system, and hands are essential. Thus, the real economy is also a collectively autocatalytic set, with "food" interpreted instead as renewable resources that feed into the economy.

In turn, economic growth of ever new goods and production capacities in the past fifty thousand years, from perhaps one thousand goods and production capacities then to perhaps ten billion now, has occurred, just as in the use of collectively autocatalytic sets not-algorithmically creating ever new niches for ever new autocatalytic sets of autocatalytic sets in an ever expanding "adjacent possible" in a biosphere of growing diversity. In the same way the growing diversity of goods and production capacities in the econosphere may well reflect the same collectively autocatalytic, non-algorithmic, niche formation. Both are likely to reflect the beginning theory of collectively autocatalytic sets, or RAFs as above, plus the non-algorithmic character by which many new goods and practical "uses" of economic "screwdrivers" emerge. Again, we are beyond entailment, for those new uses are indefinite in number and unorderable, so non-algorithmic, and not entailed. No law, it seems, entails the evolution of the economy. *A fortiori*, no law entails the evolution of culture, law, and history.

8.9 THE MATHEMATICAL THEORY OF THE SPONTANEOUS FORMATION OF COLLECTIVELY AUTOCATALYTIC SETS MAY CONSTITUTE A NEW FORM OF LAW: FORMAL CAUSE LAW

The mathematical theory of the emergence of collectively autocatalytic sets, say with respect to the origin of life, and not to an economy, posits an ensemble of chemical reaction networks each of which has molecules as substrates and products, linked by chemical reactions in a "reaction graph". In addition, in the first version of this theory, each molecule has a fixed probability of catalyzing each reaction. Which molecule actually catalyzes any given reaction is assigned totally at random. Then numerical studies and theorems show that at a sufficient diversity of molecules and reactions, so many reactions are catalyzed according to the random catalysis probability rule, that collectively autocatalytic sets "emerge" spontaneously as a literal phase transition in the specific reaction graph. Also, this theory shows that over a vast ensemble of chemical reaction graphs the same emergence would hold. Furthermore, were the constants of nature in physics changed just a bit so that chemistry changed just a bit, the same theorems would hold, so they seem not even to depend upon the physics of our specific universe.

Are the collectively autocatalytic sets that can spontaneously emerge as a phase transition, if this theory is correct, "irreducibly complex"? I think not, for although the set as a whole must come into existence as a whole, or as the expanded theory of RAFs discussed above shows, emerge by growing via slow spontaneous reactions from a food set, then become collectively autocatalytic, the ensemble of such reaction systems is "expected" to show this phase transition to collective autocatalysis. Since the transition is "expected", it is not an extremely improbable mystery, so not, as I understand the phrase, irreducibly complex.

I now ask, as hinted above, is this theory and its "law of the expected emergence of collectively autocatalytic sets" (testable experimentally by, for example using libraries of random peptides), anything like Newton's formulation? Recall, Newton mathematized Aristotle's efficient cause in his three laws of motion in differential equation form, giving the efficient cause forces between the particles and in his universal law of gravitation in differential equation form. Then, by integration, given initial and boundary conditions, we yield the deduced, hence entailed, deterministic trajectories of the particles. This mathematization of efficient cause, which for Aristotle *is* causal mechanism, *is* our mathematization of mechanism, as promulgated by Descartes in his *Res extensa*. But is the theory above of the phase transition in virtually any member of a vast ensemble of chemical reaction graphs, or more generally a set of objects, transformations among those objects, and a generalized "catalysis" of the transformations among those objects, as in an economy, a Newton-like law with efficient cause differential equation laws of motion? No, it is not. So, we might ask: what form of law is the law of the expected emergence of collectively autocatalytic sets?

Aristotle's Formal Cause concerns what it is for a statue to *be* a statue. I suggest that the theory of the emergence of collectively autocatalytic sets is a Formal Cause Law of what it is to emerge as a collectively autocatalytic set. This requires much further discussion, for it proffers a new kind of law. Note that, from the example of the economy as a collectively autocatalytic set, even the "materials" and "processes" in the emergence of the collectively autocatalytic set are "immaterial". They could be chemicals, reactions and catalysis, or input goods–output goods production functions and speeding up of production functions by some goods in the system.

Thus, we seem to have a hint of new principles and kinds of laws that may help explain the emergence of complexity in the diversifying biosphere over the Phanerozoic, and the diversifying global economy in the past 50 000 years. The answer is not to be found in some Newton-like laws of motion, but in a kind of Formal Cause Law of emergence of possible webbed collectively autocatalytic non-equilibrium systems at a sufficient diversity of "things" and "processes" by which things act on things to transform them to other things.

Could similar Formal Law principles apply to the emergence of complexity in the abiotic universe? Just maybe. I am not a physicist. But it seems that, in a fundamental sense, dissipative structures, non-equilibrium systems such as Benard cells, whirlpools, and perhaps even stars and galaxies, are linked sets of cross-coupled processes that jointly cause one another's continued co-creative, non-equilibrium existence in the universe. Let me call these abiotic physical systems "generalized autocatalytic sets". While we are used to using established physical laws to explain these phenomena, it may be that a deeper "Formal Law" theory of enough diversity of linked "things" and spontaneous and non-spontaneous "processes" will, in effect, form generalized collectively autocatalytic sets of things and processes that even create niches for one another, as do growing diversities of autocatalytic sets forming ecologies, allowing joint existence, such as the stars forming the galaxy that outlives them and gives birth to them.

ACKNOWLEDGEMENTS
This work was partially funded by the TEKES Foundation of Finland to the author as a Distinguished Finnish Professor, January 2009– December 2012.

REFERENCES
Askenasy, G. & Wagner, N. (2009). Symmetry and order in systems chemistry. *The Journal of Chemical Physics*, **130**, 164907–164911S.

Hordijk, W. & Steel, M. (2004). Detecting autocatalytic, self-sustaining sets in chemical reaction systems. *Journal of Theoretical Biology*, **227**, no. 4, 451–461.

Hordijk, W., Steel, M., & Kauffman, S. (2012). The structure of autocatalytic sets: evolvability, enablement and emergence. *Acta Biotheoretica*, **60**, 379–392. *Physics ArXiv* 1205.0584v2.

Kant, I. (2000). *The Critique of Judgment*. Translated by J. H. Bernard. New York: Prometheus Books.

Kauffman, S. (1969). Metabolic stability and epigenesis in randomly constructed genetic nets. *Journal of Theoretical Biology*, **22**, 437–467.

Kauffman, S. (1971). Cellular homeostasis, epigenesis and replication in randomly aggregated macromolecular systems. *Journal of Cybernetics*, **1**, 71–96.

Kauffman, S. (1986). Autocatalytic sets of proteins. *Journal of Theoretical Biology*, **119**, 1–24.

Kauffman, S. (1993). *The Origins of Order: Self-organization and Selection in Evolution*. New York: Oxford University Press.

Kauffman, S. (2000). *Investigations*. New York: Oxford University Press.

Kauffman, S. & Clayton, P. (2006). On emergence, agency, and organization. *Philosophy and Biology*, **21**, 501–521.

Lam, M. & Joyce, G. (2009). Autocatalytic aptazymes: ligand-dependent exponential amplification of RNA. *Nature Biotechnology*, **37**, 288–292.

Longo, G., Montevil, M., & Kauffman, S. (2012). No entailing laws, but enablement in the evolution of the biosphere. In *Proceedings of the Fourteenth*

International Conference on Genetic and Evolutionary Computation Conference Companion, pp. 1379–1392. doi 10.1145/2330784/2330946, *Physics ArXiv* 1201.2069v1 (Jan. 10, 2012), pp. 1–19.

Luisi, P. L., Mavelli, F., Rasi, S., & Stano, P. (2004). A possible route to prebiotic vesicle reproduction. *Artifical Life*, **10**, no. 3, 297–308.

Monod, J. (1995). *Chance and Necessity: an Essay on the Natural Philosophy of Modern Biology*. New York: Alfred A. Knopf.

Mossel, E. & Steel, M. (2005). Random biochemical networks: the probability of self-sustaining autocatalysis. *Journal of Theoretical Biology*, **233**, 3, 327–336.

Piel, K. T. (2012). Emotion: a self-regulatory sense. *Biophysical Psychological Review*, **2** (in press, www.emotionalsentience.com).

von Kiedrowski, G. (1986). A self-replicating hexadesoxynucleotide. *Angewandte Chemie*. International Edition in English, **25**, 982–912.

9 Emergent order in processes: the interplay of complexity, robustness, correlation, and hierarchy in the biosphere

Eric Smith

A large part of the structure that is recognized and understood in the vacuum (Weinberg, 1995; Weinberg, 1996) and in condensed matter (Ma, 1976; Mahan, 1990; Goldenfeld, 1992) is the result of a hierarchy of phase transitions in either quantum-mechanical or thermodynamic systems. Each phase transition creates a form of *long-range order* among components of the system that in the absence of the phase transition could fluctuate independently. To the extent that the phase transitions are nested or sequential in order of decreasing energy or increasing spatial scale, both of which correlate with increasing age of the universe, the progression through the sequence and the accretion of additional forms of long-range order constitute a progression of complexity.

In this chapter I argue that phase transition continues to be the appropriate paradigm in which to understand the emergence of the biosphere on Earth, and that at least some universal patterns in life should be understood as what are called the *order parameters* (Goldenfeld, 1992) of one or more such transitions. The phase transitions that formed the biosphere, however, are *dynamical* transitions rather than equilibrium transitions, and this distinction requires a different class of thermodynamic descriptions and leads to some striking differences in gross phenomenology from what has become familiar from equilibrium systems.

Complexity and the Arrow of Time, ed. Charles H. Lineweaver, Paul C. W. Davies and Michael Ruse. Published by Cambridge University Press.
© Cambridge University Press 2013.

I wish to explain six ideas and argue that they give a coherent framework for thinking about the emergence of many forms of biological complexity.

(1) Biological complexity incorporates notions of description length, memory, and robustness, all of which are naturally ensemble-derived concepts and therefore motivate the use of entropy and thermodynamic descriptions.

(2) An approach to biological complexity that emphasizes universality, stability, and predictability can suggest specific stages in the emergence of the biosphere and candidates for the order parameters of various transitions.

(3) The entropy description for biology, however, is inherently a description of processes and not merely of states. Therefore the character of the entropy functions that result, and important qualitative relations between the persistence of patterns and the persistence of entities, are different for the biosphere from forms that have become familiar from equilibrium systems.

(4) In general the role of entities in maintaining biological order is different from the role of entities in maintaining equilibrium order. In equilibrium systems robust patterns are realized by robust entities; in biology, robust patterns are maintained by ephemeral entities at many scales, and the most robust patterns (in core metabolism) are carried by the most transient entities (small metabolites).

(5) The chemical regularities of life that are its most robust feature do not seem to depend on aspects of organization at the individual level or on individuality as a concept, in striking contrast to Mendelian/Darwinian evolutionary dynamics in which individuality is fundamental. In relation to large-scale innovations in chemistry, individuality appears as a heterogeneous and derived property that has emerged in several forms in living systems, and brought Mendelian/Darwinian dynamics into existence as only one source of order within a larger universe of Markovian stochastic processes.

(6) Attempts to define life that descend from the Darwinian emphasis on replication and selection treat "life" as a predicate or property of entities; because entities play a subordinate role in the phase transition paradigm, this traditional form of definition can no longer be regarded as fundamental. I argue that the nature of the living state cannot be conflated with the nature of individuality or with properties of individuals;

a characterization of life in terms of its distinctive chemistry, for which the natural scale of aggregation is the biosphere as a whole, is a better foundation on which to unify many aspects of biological order.

In Section 9.2 I summarize core concepts from thermodynamics that frame our current understanding of hierarchical complexity in the states of matter, and show how they extend to dynamics and how dynamical order can differ from order in equilibrium. Section 9.3 reviews the traditional use of equilibrium entropy in discussions of life and explains why it is inadequate. Section 9.4 "starts over" to build a non-equilibrium thermodynamic description, beginning with empirical evidence for the appropriate kinds of regularity and proposing an interpretation of these in thermodynamic terms.

9.1 ENSEMBLES, THERMODYNAMIC DESCRIPTIONS, AND ORDER

The regularities of biological complexity that I wish to capture are those that appear as universal or convergent tendencies in all living systems, but which may not be tied to any single species, individual, pathway, molecule, or any other particular entity or event. The problem of extracting such regularities of a distribution which may not be present in any sample, and expressing them formally as physical laws, is the domain of thermodynamics. Thermodynamics is an *ensemble-based* approach, and its methods of representing order are drawn from statistical inference and the laws of large numbers.

I therefore review, firstly, the ensembles that are inherent in discussions of complexity. These include ensembles of alternatives behind the concept of the *information* in a complex state, the role of *memory* as a source of complexity in dynamics, and the importance of *robustness* against disturbances as a precondition for memory and complexity in material systems.

I then review the way *order* is represented in thermodynamic descriptions. This discussion uses a minimum of formalism, but a careful statement of concepts is essential to frame the later, more speculative interpretation of chemical flux as an order parameter of the dynamical phase transition that defines the nature of the living state.

9.1.1 Why complexity and emergence are ensemble notions

Many proposed complexity measures (Rissanen, 1989; Gell-Mann & Lloyd, 1996; Chaitin, 1990; Li & Vitányi, 2008) are linked to the notion that the observed instance (the entity or pattern which "is complex") is one form out of a variety of alternatives that might have been seen instead. The possibility of alternatives, like alternative values for a sample from a distribution, implicitly or explicitly situates complex patterns in the context of ensembles (Gell-Mann, 1994; Gell-Mann & Lloyd, 1996). The alternatives may be realizable, in which case complexity reflects genuine ambiguity in dynamics, or they may be counterfactuals that were not ruled out *a priori*, in which case complexity reflects our incomplete understanding of the restrictions on the generating process. For physical phenomena whose dynamics admit highly degenerate and thus hard-to-predict outcomes, such as glasses (Mézard *et al.*, 1984; Sherrington, 2010), the ensemble is a set of dynamically *ex*-ante-realizable states. It has sometimes been thought that for the alternative case, in which complexity is a property of a restrictive generating process that does not permit a range of outcomes, such measures are not ensemble-based. In particular, the Kolmogorov–Chaitin *algorithmic information content* (Kolmogorov, 1965; Chaitin, 1966; Li & Vitányi, 2008)[1] was meant as a measure of complexity definable for single instances (such as bit strings). However, as Rissanen (1989) has pointed out, the general-purpose computer used to define Kolmogorov's algorithmic complexity measure already makes reference to an ensemble, which is the set of executable programs all of which the computer will accept as input.

---

[1] The algorithmic information content, or AIC, is the length of the shortest program for a universal computer that will print the string in question and then halt. In general it is not computable (Chaitin, 1990), and is defined only up to an offset related to the choice of computer, and so even for application to individual strings, it is often most useful for the scaling it displays over an ensemble of structurally related strings.

In cases where unpredictable forms of order are generated by dynamics, the ensemble-notion of complexity is related to the concept of *memory*. Operationally speaking, what the particularities of each instance "remember" is something about the way the system state was constructed.

Thinking of complex systems as carrying "memories" requires us to recognize that any particular memory will have a finite characteristic lifetime and therefore limited *robustness*. If the other states in an ensemble of *ex ante*-realizable possibilities are truly "alternatives" to the state seen, then they could have been produced by other histories equivalent to the one that actually transpired, but differing in freely interchangeable details. (It is through the notion of memory about interchangeable details that complexity is connected to historical contingency (Gould, 1989, 2002).) Real systems are always made up of fluctuating components. As a consequence of fluctuations in the matter and energy that make up the physical world, the implicit ensemble that makes a system complex ultimately becomes, on sufficiently long timescales, a dynamically sampled ensemble that erases memory.

For these reasons, formal and especially quantitative treatments of complexity will inherently require several related ensemble-derived notions: of the information needed to restrict samples or generating processes, of memory, and of robustness. These are the concepts underlying thermodynamics as a particular application (Jaynes, 1957a, 1957b) of information theory (Cover & Thomas, 1991). Differences between equilibrium and non-equilibrium applications are secondary to these basic ideas. Therefore I will briefly summarize the ways thermodynamics deals with the problem of identifying essential order in systems that are also subject to sample fluctuation.

9.1.2 Fluctuations, sufficient statistics, and order parameters

Thermodynamics allows us to address the fact that, in most large systems, samples of system states reflect both a central tendency that will be common to all typical samples, and non-essential fluctuations of any given sample away from this central tendency. The two problems that thermodynamics must solve are identification of likely functional classes for generating processes, and the estimation of the parameter values that govern any particular system of interest. It is the structure and values of the parameters characterizing the central tendency that we associate with "laws" for the complex system.

The functional classes that define thermodynamic limits are those given by *large-deviations scaling* (Touchette, 2009), a particular application of the laws of large numbers (Ellis, 1985). In a system that converges towards a large-deviations scaling limit, the probability distributions for fluctuations take on a simple exponential form, in which *scale separates from structure*. More specifically: if some number N gives a measure of a system's size, such as its mass, energy, volume, current flux, etc., and if, for a particular class of observed values n, a quantity $x = n/N$ describes the structure of a fluctuation that can exist at many scales, then the log-probability of observing n in the large-deviations limit takes the form

$$\log p(n) \approx f_{\text{scale}}(N)s(x), \tag{9.1}$$

where the $\approx$ indicates that these terms capture the leading exponential behavior of the probability. $f_{\text{scale}}(N)$ describes a common exponential suppression of fluctuations in large systems purely as a consequence of their change in scale, *without regard to which fluctuation-form we consider*, while $s(x)$ – known as the *rate function* or *entropy* of the large-deviations limit – distinguishes likelihoods of different kinds of fluctuation *without regard to the system size*. Systems whose fluctuation probabilities take this form are said to have a *thermodynamic limit*.

The restriction to exponential probability distributions for samples of a system's state gives an enormous simplification of the description of non-essential fluctuations. But to characterize the law-like property of a thermodynamic system, we must still identify the central tendency from which deviations are suppressed. The quantities that capture central tendencies of samples, from which we estimate the parameters of the underlying generating process, are called *sufficient statistics*. Suppose, for example, that observed states of some system might (*a priori*) be generated by several processes that we can distinguish by a model-parameter θ, and the lawful description of the system derives from the estimation of θ. If x is a (generally complicated) property we observe about the system's state, some particular (and generally simpler) function $T(x)$ is *sufficient for* θ if any information about the generating process (θ) in the sample x is contained in $T(x)$.

The entropy $s(x)$, and any scale-independent parameters on which it depends, are sufficient statistics for the mean behavior and macroscopic fluctuations in a thermodynamic process; all further details of fluctuations about these background quantities, such as the number of components that they involve or the absolute magnitude of deviation probabilities, depend only on $s(x)$ and the scale factor $f_{scale}(N)$.

In thermodynamics, the *order parameter* (Ma, 1976; Goldenfeld, 1992) is a sufficient statistic for the regularities or law-like behavior in systems with both a central tendency and non-essential fluctuations. New forms of order are said to *emerge* in a thermodynamic system, as a result of changes in its boundary conditions, when some order parameter goes from a zero value reflecting a less-constrained set of fluctuations to a non-zero value that reflects a more-constrained set of fluctuations. An example is the transition from a melted state, in which atoms or molecules can rotate independently in any direction, to a frozen state, in which they share a tendency to some common orientation and only show suppressed fluctuations about that common orientation. The emergence of successive layers of order in a system, through processes such as phase transitions, leads to sufficient statistics that require more and more elaborate description, and which give more and more information about the likely states of samples. The accretion of layers of order in a system that make its sufficient statistics more elaborate to specify, and its ensemble of states more constrained, is the framework by which thermodynamics has characterized the emergence of hierarchical complexity in the universe's states of matter.

Nothing in the foregoing characterization of complexity and thermodynamics presumes equilibrium. If the biosphere presents evidence of convergence of living systems toward a central tendency that reflects hierarchical or increasing complexity, the same problems of distinguishing law from non-essential fluctuation arise, and the same tools are suggested to address those problems. However, much of our thermodynamic intuition has derived from equilibrium mechanical systems, and *the intuition may not generalize*, so it is important to note ways in which dynamical order and complexity measures may be different and new.

9.1.3 Irreversibility, non-equilibrium entropies, and ensembles of histories

The thermodynamic entropy function for a system gives the dependence of the system's distribution of states on its boundary conditions. For equilibrium systems whose distributions are not only time-stationary but *time-reversible* (Onsager, 1931a, 1931b), the thermodynamic entropy can only depend on properties that are preserved under time-reversal of dynamics. Entropy functions for systems driven away from equilibrium will therefore generally be different functions from the equilibrium entropy, even though they are defined by the same principles (Smith, 2011). If a system's environment drives the system away from equilibrium and induces currents and possibly other patterns, the environment *constrains* the system more, we therefore *know* more about the system, and samples of the system's states will be more *limited*, than those for a system with the same contents but permitted to relax to equilibrium.

Figure 9.1 uses the familiar example of a *free energy landscape* to illustrate some ways that ensembles and entropies can differ at and away from equilibrium.[2] Such a landscape might describe a space

---

[2] For a more thorough derivation starting from this figure, see Smith (2011).

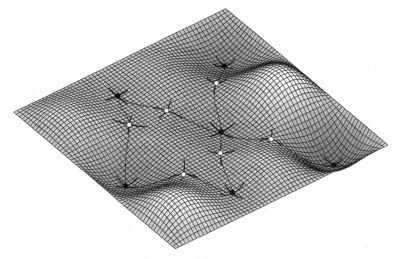

FIGURE 9.1 A two-dimensional free-energy landscape, on which the low-temperature statistical mechanics is dominated by neighborhoods of a small network of paths. Local minima are attracting fixed points (filled dots), near which the system spends most of its time. When transitions between minima occur, the conditional probability for their trajectories is dominated by a small collection of escape paths (lines), which pass with high probability through the saddle points (open dots) between the initiating and terminating minima.

of possible chemical species and reactions, for example. The free energy landscape has two important classes of fixed points for the small-scale dynamics: attracting fixed points (filled dots) and saddle points (open dots), which are attracting in all but one direction but repelling in the final direction. In the common thermodynamic limit, the system spends almost all its time in small neighborhoods of the attracting fixed points, so almost all other details of the landscape can be ignored. However, the *conditional* distribution for histories, conditioned on the existence of some transition, are dominated by small neighborhoods of the saddle points, and by most-likely paths that connect them to the attracting fixed points, making a *heteroclinic network* (Gluckheimer & Holmes, 1988). The relative heights of the saddle points above the attracting fixed points determine the rates of transition (Glasstone *et al.*, 1941). If transitions through each saddle are allowed to come into detailed balance (Gray, 1994), so that currents in both directions are equalized for every transition, the relative heights of the attracting fixed points determine their occupation probabilities.

The entropy function for an equilibrium process on such a landscape, which presumes a sufficiently long waiting time that reaction rates have all come into detailed balance, depends only on neighborhoods of the attracting fixed points. The entropy function for a process that is out of equilibrium on the same landscape will generally depend on neighborhoods of the saddle points as well.

The increase in the number of properties that must be specified to describe rates makes the non-equilibrium entropy a more complicated function than the equilibrium entropy, and one that can depend on *currents* through the saddle points (Smith, 2005, 2011) as well as on time-reversal-invariant properties such as total matter and energy contents. More importantly, the system could be driven out of equilibrium by placing sources and sinks at several different points on the landscape, so the *network topology* as well as the properties of saddle points may affect the non-equilibrium entropy. Whole domains of configuration space may be cut off, because the reaction rates make them unreachable between the time that matter or energy enter at a particular source and the time they exit through some drain. The properties of rates in such an ensemble are collectively termed its *kinetics*; the opening or closing of large domains in configuration space is a key consequence of the evolved control over kinetics in living systems.

For some problems, the additional knowledge and constraint imposed by kinetics can be incorporated by fairly straightforward extensions of equilibrium entropy functions (Smith, 2005), while for others one must shift the entire ensemble description from considering sets of states to considering sets of histories, leading to an entropy function of qualitatively different structure (Smith, 2011).

In equilibrium thermodynamics, one identifies the most-likely state (the central tendency, and the carrier of the system's law-like behavior) by maximizing the entropy of a system in the context of its surroundings (Fermi, 1956). For equilibrium systems, this entropy is generally a volume-integral of a local *density function* called the *specific entropy* (Kittel & Kroemer, 1980). The system states can be identified from maxima of a local function because the system degrees of freedom find their most-likely distribution independently at all points of space. The counterpart to the equilibrium entropy for dynamical ensembles is a thermodynamic version of the Lagrange–Hamilton action functional (Goldstein *et al.*, 2001) known as an *effective action* (Smith, 1998, 1999, 2011). Finding minima of the effective action is on one hand the non-equilibrium generalization of finding maxima of entropy (minima of free energy) (Freidlin & Wentzell, 1998); on the other hand it is the thermodynamic generalization of finding macroscopic mechanical trajectories. A remarkable fact is that effective actions constructed for non-equilibrium ensembles are generally integrals of *non-local* densities (Bertini *et al.*, 2009; Smith, 2011), reflecting the presence of long-range correlations among the distributions of spatially separated degrees of freedom.

Non-local interactions, across time, space, or system components, can greatly increase the difficulty of elaborating ordered states. Perhaps for this reason (as well as for others particular to chemistry), biological order is categorically more complex than order in non-living physical systems. However, some of the complexity of life may be a simple consequence of the transition from equilibrium to dynamics.

9.2 COMPLEX ORDER IN THE BIOSPHERE

9.2.1 Discussions of entropy in biology

Discussions of entropy in relation to biological order are not new. However, so far they have not led to many useful predictions about the emergence, organization, or robustness of living systems. I will first explain why these traditional discussions can only have limited scope, and then argue that the problem is not with thermodynamic descriptions, but with the limited information available from the equilibrium entropy.

Boltzmann, Schrödinger, and contemporary

Boltzmann, in formulating the concept of entropy, was already well aware that his arguments for a tendency toward disorder would need to be consistent with the apparently "self-generated" order of life. He also understood why there is no conflict (Boltzmann, 1905): systems in nature are connected, and the increase of entropy refers only to the connected whole, not to each of its components separately. Schrödinger (1992), Brillouin (2004), and Quastler (1964) have all reiterated this point. All these arguments refer to the *equilibrium* entropy function. Equilibrium entropy characterizes the least-constrained distribution with a given content of matter, energy, volume, etc., ensuring that the Boltzmann limit on the decrease of entropy within subsystems is always an ultimate bound. However, for non-equilibrium systems which (as we have noted) are subject to more constraints than those at equilibrium, the equilibrium ensemble may be too permissive, rendering bounds derived from the equilibrium entropy *loose* and therefore uninformative.

Prigigone (Kondepudi & Prigogine, 1998) has derived laws of motion for some irreversible systems from rates of change of the equilibrium entropy. However, I have shown (Smith, 2005) how systems which have simple non-equilibrium entropy functions may project onto the equilibrium entropy in a way that loses the actual constraints responsible for dynamics. A systematic treatment derived from entropies of histories was recognized by Kolmogorov and Sinai (Kolmogorov, 1958; Sinai, 1959), and subsequently elaborated by Jaynes (1980) and others (Ghosh *et al.*, 2006; Stock *et al.*, 2009; Wu *et al.*, 2009). This treatment, which *keeps* the effects of kinetics, can be used to show which aspects of dynamics remain ambiguous from properties of the equilibrium entropy alone (Smith, 2011).

Therefore, while entropy has been discussed in connection with life for more than a century, the *relevant* measures of entropy to address the emergence and complexity of life were not part of that discussion. In Boltzmann's or even in Schrödinger's time, it would have been difficult to explain what is left out, but with the familiarity of thermodynamic machines in modern life, the differences have become much easier to understand.

A parable of refrigerators

Equilibrium thermodynamics, by telling us how refrigerators work, entails the more interesting result that *refrigerators can exist* (Fermi, 1956). Entropy can be pumped out of matter at one place, in the form of heat, if at least that much entropy is delivered or produced somewhere else in the system (at a power plant). This is the analog (Smith, 2003, 2008a, 2008b) to Boltzmann's observation that life is not in conflict with the second law of thermodynamics. However, that knowledge alone is not enough to explain *why* refrigerators were invented, why new ones are built as old ones wear out, or how long such machines are likely to persist in our everyday life and experience.

To answer those questions we must study the paths that lead to the invention and manufacture of refrigerators, and the context of larger systems that produce and use them. The paths may seem very particular, and in the case of technologies, the number of levels we need to consider may be daunting. But, even in this case, such paths may not be arbitrary. They may be the least-improbable routes to a particular pattern: the probability for the emergence of refrigerators on Earth may be dominated by a path that goes through the emergence of life, then cognition, then language and culture, then science and technology. Herbert Simon argues (Simon 1962, 1973) that this reliance on intermediate levels of assembly will generally characterize the most-probable paths to entities at high levels of hierarchical complex systems. More formally, we could say that the conditional probability for histories of the Earth and everything in it, *conditioned on the presence of refrigerators*, is strongly dominated by histories in which technological humans were the agents of their invention and production. This is a very fancy version of the result shown in the last section for free-energy landscapes: the conditional probability for histories of a system over a free energy landscape, conditioned on the existence of one or more transitions, is dominated by small neighborhoods of the saddle points through which those transitions are most accessible.

The parable of refrigerators emphasizes that the emergence of life requires a thermodynamics of processes, and that these processes can be richly and hierarchically structured. The role of human intelligence in the parable probably makes it hopeless to formalize. Fortunately, for biology in contrast to human technology, there is evidence that the number of levels that must be considered to reconstruct the emergence of the biosphere is not intractably large; for the lowest levels of emergence of chemical order, many details of biological organization appear to be secondary.

Core metabolism as evidence for lawfulness and sequence in the emergence of life

A small and barely-diversified network of pathways is responsible for the conversion of inorganic carbon and nitrogen sources into biological molecules in all ecosystems on Earth (Morowitz, 1992; Smith & Morowitz, 2004; Braakman & Smith, 2012). At the core of this network are the molecules and reactions of the citric-acid cycle and some reactions in the folate pathway of one-carbon metabolism (Braakman & Smith, 2012). The citric-acid cycle reactions may act directly to fix carbon (Hügler *et al.*, 2005; Braakman & Smith, 2013), or they may play the role of so-called "anaplerotic" pathways (Lengeler *et al.*, 1999) that connect some other carbon fixation pathway (Berg *et al.*, 2010; Hügler & Seivert, 2011) to the precursors of biomass. Their function in either role is highly conserved, and four or five intermediates of the citric-acid cycle are the precursors for all biomolecules (Morowitz, 1992). A similarly invariant role is played by the pentose-phosphate pathway to sugars including ribose (Stryer, 1981), a collection of tightly homologous pathways on folates and pterins that mediate one-carbon chemistry (MacKenzie, 1984; Maden, 2000) and in some organisms fix carbon (Ljungdahl *et al.*, 1965; Martin, 2008), and the early reactions in the pathways that form amino acids and nucleotides from keto-acid precursors (Copley *et al.*, 2005).

The molecules and reactions in this core network are more conserved than the genes for the enzymes that support them (Doolittle, 2000). The natural boundaries of modules in the network – which also correspond to distinctive types of chemical reactions – can be seen to be recapitulated in structure at many higher levels in metabolism, including the transition from organic to phosphate chemistry patterns in the stabilizing selection of catalytic residues, and the retention or convergence of orthologous families of enzymes (Braakman & Smith, 2013).

Following Morowitz (1992) and several others (Wächtershäuser, 1990; Russell & Martin, 2004) (with some difference of detail) we have argued (Copley *et al.*, 2007; Morowitz & Smith, 2007; Braakman & Smith, 2013) that these core pathways are good candidates for the first layer in emergent life, and the template that constrained many of its later features. Even prebiotically they required catalytic support, but limited experimental work in this area suggests that minerals are plausible catalysts for relevant reactions (Hüber & Wächtershàuser, 2000; Cody *et al.*, 2001). An emergence of the biosphere through small-molecule organometallic chemistry, unlike the parable of the refrigerator, relies only on a denumerable (if still large) space of possible constituents and a limited set of plausible transitions.

Our arguments up to now have mostly marshaled evidence for antiquity and universality but have not explicitly used an interpretive mathematical framework. In Section 9.4 I offer an interpretation within the framework of phase transition summarized in Section 9.2.

9.2.2 Patterns and entities in the biosphere
The oldest fossils on Earth

The conserved core network may constitute the oldest fossil on Earth. It appears to antedate the split of the bacterial and archaeal domains (Martin & Russell, 2006; Braakman & Smith, 2012). While it seems likely that the last common ancestor of all modern lineages was a kind of cellular organism (Peretó *et al.*, 2004), enough elements of metabolism are more common than major cellular systems such as DNA replication machinery, phospholipid chirality, and cell-wall architecture, that cogent arguments have been made for a last common ancestor that was a non-cellular, virus-like, RNA organism precursor to fully-formed modern cells (Martin & Russell, 2003; Koonin & Martin, 2005). Relative to these early stages, the oldest preserved rocks showing evidence of biological mineral modification are believed to have been the work of cells possessing an essentially modern inventory of catalysts and cofactors. The core network thus appears to be a biological pattern showing *no alternatives* and *indefinite persistence.*

The antiquity of core metabolism and the fidelity with which it appears to have been preserved highlight a property of biological order that stands in sharp contrast to ordered states in equilibrium thermodynamics. Core metabolism is instantiated by populations of small molecules that may have lifetimes from less than a second to hours or days. The molecules are coordinated in cellular processes that undergo constant regulatory adjustment and re-arrangement on the scales of cellular lifetimes (hours to weeks). Species-specific genotypes and ecological relations that distinguish cells are maintained on timescales that may be as short as decades (some bacterial strains) or as long as millions of years (common vertebrate species). At every level of biological organization, specific ordered states – molecules, cells, species, ecosystems – turn over rapidly compared to the persistence time of the patterns they instantiate and maintain.

The contrast with persistence in equilibrium systems, such as fossil minerals, is categorical. The pattern that is distinctive of quartz is a particular crystallographic unit-cell structure. The realization of the crystalline form in matter persists because the *physical piece of quartz* is robust against disruption. Yet all known quartz minerals on Earth have been reworked geologically since the age of the original core metabolism, which has been maintained by steadily overturning populations of small molecules and cells.

The systematic dependence of durable patterns on ephemeral entities may be the most distinctive difference between biological order (or non-equilibrium order more generally) and equilibrium order. It is a characteristic of all levels of interaction in the biosphere. Cell homeostasis results from constant turnover of molecular populations. Species identities are preserved by populations of continuously regenerating and dying cells. Ecological community structures are maintained by a flow-through of organisms and species, yet they can preserve network-topological features for hundreds of millions of years (Dunne *et al.*, 2008). Most entities in biology play both roles at once: the cell is a durable pattern with respect to its molecular turnover, but it is the ephemeral carrier of the species' genome and its ecological position. Most striking, though, is that the most durable single feature of life (core metabolism) is the immediate expression of collections of its shortest-lived entities (small metabolites).

A meaningful sense of pattern maintenance
To show that this is a meaningful comparison, it is important to check that *potential patterns* in one domain are not compared to *physical instantiations of patterns* in another.

A potential pattern is the point-group symmetry of a crystalline unit cell, or the unique role of specific core metabolites and pathways in the energy flows of life. The unit cell exists as an abstraction even if no crystals exist that instantiate it; it could be derived from the quantum mechanics of the constituent atoms. The role of a core metabolite exists as a similar abstraction; if a predictive account of the spontaneous emergence of metabolism existed, its role could be derived from physical chemistry and network and energetic context.

The importance of the real lithosphere and the real biosphere is that crystals exist which realize those unit-cell symmetries, and organisms exist which catalyze the metabolic pathways. They interact and imprint their patterns on other material systems. The durability of this imprint is a consequence of collective effects in equilibrium phase-transition theory, or of dynamical collective effects in the biosphere (Goldenfeld & Woese, 2011).

It is remarkable that the full machinery of life is recruited to actively maintain the universal core pathways, even though we have no evidence that there is any alternative to them, or that life is fragile because they may be lost. We tend to think of natural selection in relation to adaptation, but a large background of normalizing selection is constantly at work preserving the fidelity of enzymes and their integration into simple metabolic networks. A full understanding of the stability of the biosphere will require us to explain how so many levels can be reinforced by their support of such distinctively biological, but otherwise universal, forms of order as core metabolism, and also why so many levels seem to be required.

9.3 THE NATURE OF THE LIVING STATE

I now argue for a thermodynamic view of the emergence, organization, and robustness of life as a phenomenon in chemistry, and show that this conceptualization greatly reduces the salience of individuality and requires a re-interpretation of its meaning and its role.

9.3.1 Universality, robustness, and predictability bypass individuals and favor ecosystems

An emphasis on regularity that does not begin with individuality or individuals

The main difference in my approach to the nature of the living state here, from most approaches in biology, is that I do not presume that the most important questions are "who are the entities and what do the entities do?". From antiquity to Darwin's time this seemed the obvious question: both unproblematic in its conception and self-evident in its importance. Across biology within the twentieth century, however, both of these presumptions have been eroded. In development (Buss, 2007), ecology (Brown *et al.*, 2004; Dunne *et al.*, 2008; Odling-Smee *et al.*, 2003), epidemiology, and evolutionary dynamics (Krakauer *et al.*, 2011), the identification of entities has become more ambiguous and the importance of organizational patterns that are not naturally characterized as entities (Hamilton, 1964a, 1964b; Lewontin, 1974) has become more apparent.

I therefore began with an orthogonal set of questions, appropriate to the study of any complex system subject to both law and accident: What is universal and what is variable? What has (apparently) emerged in a deterministic sequence and what has fluctuated? What appears predictable and what appears historically contingent?

Metabolism at the ecosystem level becomes
the reliable observation
In this sequence of questions, core biochemistry consisting of carbon fixation and intermediary metabolism emerges as the most universal trait and the one whose formation seems most progressive. The preservation of the core network does not seem to have relied on the preservation of species lineages, although some of its diversification may have followed (or led to) diversification of the major clades (Braakman & Smith, 2012). Instead, ecosystems emerge (Morowitz & Smith, 2007; Smith & Morowitz, 2010) as the level of organization more closely associated with the universality of biochemistry than organisms. Indeed, the conservation of the network is often only visible at the level of the trophically complete ecosystem; the division and diversification in the use of particular pathway segments among species is extensive (Rodrigues & Wagner, 2009). Yet the paths of biosynthesis of most of the essential biomolecules – even when they are interrupted by catabolic reversals or detour through salvage pathways, and even when they proceed through several species on the way to completion – are *in toto* highly conservative, and more so in the core than in the periphery of the network (Csete & Doyle, 2004; Riehl *et al.*, 2010).

The nature and role of individuality reconsidered
Not only do particular kinds of individual show no special significance from this statistical perspective; the nature of individuality emerges

more clearly as a concept distinct from the property of participation in the biosphere's chemistry. Individuality is not a unitary notion represented in a single form. It is a complex property of both local interactions and covariance in population dynamics. Briefly: (1) the entities that we regard as individual have strongly non-linear spatiotemporal boundaries that make their interactions in the world "granular", leading to the distinctions between self and other or system and environment that make Darwinian competition possible, and the temporal generation structure that creates the notion of "replication"; (2) biological individuals are more than physical particles; they have multiple distinct traits that nonetheless reproduce or die as a package, a property of covariance that might be called "shared fate". Many kinds of living order possess these two properties of granularity and shared fate. DNA or RNA preserve a partial autonomy from the metabolizing cell in the ecology of viruses (Claverie, 2006; Forterre, 2010); host-cell and viral genomes compete for control within an organism; the soma is subordinated to the reproductive interest of the germ line in multicellular organisms (Buss, 2007); higher-order groups are eligible units of selection depending on their relatedness and environment (Hamilton, 1970; Frank, 1997). Each of these forms of individuality appears as a particular derived form of functional organization, and all of them are at work concurrently in any typical ecosystem.

If individuality is thus secondary and emergent, and if the modularity of biochemistry and physiology in organisms does follow outlines prefigured in core metabolism, and if the ecosystem is the natural level of aggregation at which core metabolism displays universal pathways and motifs, then the universality of the network, which does not depend on species, seems to provide a prior constraint to the small functional variations that are created by the elaboration of species and ecological complexity. It may be more appropriate to assign the *variations* in the use of core metabolism to the interaction of species lineages within trophic ecological communities, than to regard individuality as a likely source of the very deep chemical regularities of life.

The presumption of individuality in definitions of life

The common formulation "Life is that which replicates and evolves" (Nowak & Ohtsuki, 2008) then becomes difficult to use as a fundamental characterization of the nature of the living state. Here "life" is treated as a property of a class of objects. But the objects are derived: replication, competition, and selection require an individual substrate and the concept of individuality grows more heterogeneous and ambiguous as our understanding of it improves. Even where individuals exist, they seem not to be fundamental: reconstructed bacterial or archaeal lineages seem superfluous to the chemistry that is the most universal and law-like feature of life. If we are to place individuality where its contribution and variability seem to require – as an emergent organizational motif within "something else" – we must have a different way of speaking about the nature of the living state.[3]

9.3.2 A chemical flux as the defining characteristic of the living state

In place of individual-based characterizations of life, I want to propose a definition in terms of the unique capacity of life to access certain domains of organic chemistry, and the whole biosphere as the level of aggregation that implements and thus defines the living state. The distinction between entities that participate in the biosphere as its components and the carriers of its functions, and those that do not, replaces the older formulation "'Life' *is* 'that which'....", and serve the same function but with greater flexibility. The transitory and fungible role of almost all entities in this way of speaking would have appeared too loose to carry meaning, in a pre-thermodynamic era where distributional properties of relations were not recognized as carriers of physical law. In a post-thermodynamic age which recognizes

---

[3] Note, however, that if we do not presume "life" is a property of objects, traditionally problematic formulations such as "are viruses alive?" cease to appear as paradoxes.

that all previously-presumed objects *are defined from* distributional properties of relations,[4] a characterization of life in the same terms seems not only possible but inevitable.

The significance of the distribution of matter and reaction flux through domains of organic chemistry

The chemical characterization of the living state begins with a comparison of the Earth without, and Earth with, a biosphere as an integral part of its dynamics and energetics. Consider the space of possible chemical species and chemical reactions, given the elemental composition of our planet. Of particular interest are the organic and organometallic compounds involving covalent bonds to C, H, O, N, P, and S, and transition metals (Morowitz *et al.*, 2010), which have either bond free-energies or reaction transition-state energies ΔG (relative to the chemicals in an equilibrium ensemble) much larger than (say, 50–100 times) the thermal activation energy $k_B T$. Absent a biosphere, terrestrial atoms would never (or almost-never) access this part of the chemical possibility-graph, because the Arrhenius factor $e^{-\Delta G/k_B T}$ suppresses reaction rates and fluctuation probabilities from an equilibrium ensemble.

Compare this case to the Earth with a biosphere. Most of the chemical graph remains unoccupied by terrestrial atoms, but a small subset – the part that can be built by phosphate dehydration or physical assembly from about 125 particular small core metabolites (Srinivasan & Morowitz, 2009b, 2009a) – has a continuous current of gigatons of carbon (and comparable amounts of O, N, and H) per year flowing through it. The biosphere is the collection of processes that concentrate a subset of geochemical electron-potential (redox) energy or electromagnetic (solar flux) energy and cause/enable it to flow through the chemicals and reactions of this very distinctive subgraph

---

[4] This is the central message of the renormalization group and effective field theory (Wilson & Kogut, 1974; Goldenfeld, 1992; Weinberg, 1995, 1996), and it is the most important conceptual revision in 20th century physics.

of organometallic chemistry. Some forms of organization that mediate this flow are simply harnessed from near-equilibrium geochemical ensembles (the water solvent, phase-separated mixtures (Luisi, 2006), or metal-sulfide centers resembling minerals (Russell & Hall, 1997, 2006), are well-known examples). Other forms of organization, however, would never be sampled in an equilibrium ensemble; if dynamics cannot adequately maintain and renew them, they cannot persist. The highly selected (Copley *et al.*, 2010) forms of order that can be self-maintained (as a system) and that contribute to the access of a unique chemical and energetic space define the nature of the living state.

The flux through core metabolism is to be interpreted as the order parameter of a non-equilibrium phase transition that occurred in geochemistry as the first stage (or stages) in the emergence of the biosphere. Because all metabolites flow through the core, the flux in the core is a *function of* all secondary fluxes, defined by the rules of chemical stoichiometry. As a function of the whole collection of fluxes that constitute a sample of the state of the biosphere at any moment, it satisfies the definition of a summary statistic. (Demonstrating that it is a sufficient statistic is the entire science problem of validating a metabolic origin of life.) The specification of the aggregated network of core pathways through which this flux passes, including the relative degrees of universality or context-dependence of different pathways and sub-networks, is the refinement of the summary statistic of aggregate flux to a more detailed description of the distribution of eligible metabolic forms. A derivation of higher-order structures as functional components in the system whose feedback maintains the core flux – showing that such motifs as molecular replication, cellular compartmentalization of metabolism, or the modularization of biological energy systems both contribute to maintaining the core flux and are maintained by it through selection – further refines the dependence of restrictions of biological form on the stable order parameter. Residual variations that are truly free then define the entropy of the distribution over evolutionary histories.

Darwinian evolution is the Markov process created by the emergence of individuality

Molecular replication, cellular compartmentalization of metabolism, and other forms of granularization of dynamics and linkage of traits in populations are then proposed to have arisen through stochastic feedbacks as favored mechanisms to support metabolic flux (displacing mineral catalysis (Huber & Wächtershäuser, 2000), adsorption on sub-crustal surfaces (Wächtershäuser, 1988), or confinement in mineral foams (Russell & Martin, 2004)). The advent of these forms of individuality brings into existence Darwinian selection, as a specialized subset within the larger space of Markovian stochastic processes.

The role of individuality sets Darwinian evolution apart within the Markov processes, not only formally but in the essential use of variation and the consequent capacity to lock in accidents of history. The maintenance of standing variation which is essential to Darwinian evolution (Fisher, 2000) is not a property easily achieved by systems below the level of integrated organisms; the long-range order created by genome integration with its celebrated consequence of pervasive historical contingency (Gould, 2002) is perhaps biology's most important source of standing variation in populations occupying common environments. In the deep past when the whole genome was much less a unit of transmission due to horizontal gene transfer (Woese, 1987, 2000), evolution more closely resembled bulk-nonequilibrium thermodynamics with loosely coupled components and fewer barriers to independent optimization. Innovation-sharing led to robust convergence on the highly-optimized error-tolerant genetic code (Vetsigian *et al.*, 2006), but preserved almost no variants over long times. We have found a similar, surprisingly tree-like history of innovations in early carbon fixation networks (Braakman & Smith, 2012), which we argue is most naturally understood as a result of local adaptation by organisms sequentially populating new geochemical environments, but unable to carry any individually optimized phenotype outside its preferred environment. Pathways adapted to different environments could never combine to produce the usual reticulated tree that is found for aggregate gene phylogenies near the root (Puigbo *et al.*, 2009). This early behavior, which we interpret as a consequence of *inability* to maintain standing variation in populations in a common environment, is in stark contrast to the evolution of core carbon metabolism later during the rise of oxygen, when clades such as methylotrophic α-proteobacteria (Chistoserdova *et al.*, 2009) drew metabolic modules from branches across the tree of life that had evolved separately for more than 1.5 billion years.

9.3.3 Why chemistry gives a better foundation than Darwinism for the complexity of life

The importance of Darwinian evolution in the ongoing dynamics of the biosphere cannot be contested, nor can the developmental importance of cells and organisms, or the constraints to which ecosystems are subject as community assemblages of individual organisms. But all these concepts exist within contexts; kinetics, network topology, space-filling, or other physical influences intrude sometimes as strict constraints, sometimes as strong evolutionary convergences. The problem of choosing an essential foundation for the nature of the living state is that the other patterns of life should either follow from that foundation or be truly arbitrary. I have reviewed a few things left out by attempts to use individuality or Darwinian evolution as such a foundation, and suggested how evolution may more naturally grow out of chemical dynamics. *Many* more details about the configuration space of chemical structures, emerging from the quantum mechanics of molecular bonds, are directly relevant to the distinctive complexity of life, but further discussion becomes detailed and regrettably is beyond the scope of this chapter.

9.4 CONCLUSION

Beneath the organizational level of individuals and communities, and the adaptive dynamics of Darwinian selection, life is distinguished from non-life by a sparse and universal chemistry that appears to be associated with the biosphere as a whole, or at least the self-contained ecosystem level of aggregation. All biosynthetic fluxes pass through this core network, and multiple levels of feedback (physiological, evolutionary, ecological) contribute jointly to its maintenance. The challenge in defining the nature of the living state and forming a theory of biological complexity has been accounting for such regularities which do not seem to depend on any particular component, individual, or species, but which are maintained in essentially the same form by many distinct and complex assemblies of living systems. I have argued that such regularities are the natural domain of thermodynamics, and sketched the main concepts of a non-equilibrium thermodynamics applicable to a chemical basis of life and an information-theoretic approach to complexity.

The argument that metabolic flux is a foundation for the biosphere satisfies our intuition that the emergence of life marked a jump in the complexity of the Earth, and it provides a way to quantify *a part* of that jump. The difference between my approach to life's complexity in this chapter, and the traditional approach of natural history – cataloguing the components and their relations – is akin to the difference in definitions of complexity due to Gell-Mann and Lloyd (1996) or to Kolmogorov (1965). Kolmogorov's Algorithmic Information includes both repetitive order and order we might think of as "random"; it is largest for totally-random instances. Gell-Mann and Lloyd advocate measuring the complexity of the *regularities* in a system, and omitting a catalogue of random features. The latter approach requires making judgments about what to regard as random, so *the judge* becomes part of the complexity measure. The metabolic foundation of life is meant to be its most-robust pattern, requiring recognition by all judges. The ability to support many higher levels of living structure – even when it does not fix their particular forms – is a property of that foundation and part of the measure of its complexity. But for shorter-term or context-dependent questions, other judgments will surely be needed to add more particularities as expressions of laws in a cascade of refinement.

REFERENCES

Berg, I. A., Kockelkorn, D., Ramos-Vera, W. H., *et al.* (2010). Autotrophic carbon fixation in archaea. *Nature Rev. Microbiol.*, **8**, 447–460.

Bertini, L., De Sole, A., Gabrielli, D., Jona-Lasinio, G., & Landim, C. (2009). Towards a nonequilibrium thermodynamics: a self-contained macroscopic description of driven diffusive systems. *J. Stat. Phys.*, **135**, 857–872.

Boltzmann, L. (1905). *Populate Schriften*. Leipzig: J. A. Barth, re-issued Braunschweig: F. Vieweg, 1979.

Braakman, R. & Smith, E. (2012). The emergence and early evolution of biological carbon fixation. *PLoS Comp. Biol.*, **8**, el002455.

Braakman, R. & Smith, E. (2013). The compositional and evolutionary logic of metabolism. *Physical Biology*, **10**, 011001, doi:10.1088/1478-3975/10/1/011001.

Brillouin, L. (2004). *Science and Information Theory*, second edn. Mineola, NY: Dover Phoenix Editions.

Brown, J. H., Gillooly, J. F., Allen, A. P., Savage, V. M., & West, G. B. (2004). Toward a metabolic theory of ecology. *Ecology*, **85**, 1771–1789.

Buss, L. W. (2007). *The Evolution of Individuality*. Princeton, NJ: Princeton University Press.

Chaitin, G. J. (1966). On the length of programs for computing finite binary sequences. *J. Assoc. Comp. Machinery*, **13**, 547–569.

Chaitin, G. J. (1990). *Algorithmic Information Theory*. New York: Cambridge University Press.

Chistoserdova, L., Kalyuzhnaya, M. G., & Lidstrom, M. E. (2009). The expanding world of methylotrophic metabolism. *Ann. Rev. Microbiol.*, **63**, 477–499.

Claverie, J.-M. (2006). Viruses take center stage in cellular evolution. *Genome Biol.*, **7**, 110:1–5.

Cody, G. D., Boctor, N. Z., Hazen, R. M., Brandeis, J. A., Morowitz, H. J., & Yoder, H. S. J. (2001). Geochemical roots of autotrophic carbon fixation: hydrothermal experiments in the system citric acid, H_2O-($\pm FeS$) ($\pm NiS$). *Geochimica et Cosmochimica Ada.*, **65**, 3557–3576.

Copley, S. D., Smith, E., & Morowitz, H. J. (2005). A mechanism for the association of amino acids with their codons and the origin of the genetic code. *Proc. Nat. Acad. Set. USA*, **102**, 4442–4447.

Copley, S. D., Smith, E., & Morowitz, H. J. (2007). The origin of the RNA world: co-evolution of genes and metabolism. *Bioorganic Chemistry*, **35**, 430–443.

Copley, S. D., Smith, E., & Morowitz, H. J. (2010). The emergence of sparse metabolic networks. In M. Russel (ed.), *Abiogenesis and the origins of life*. Cambridge, MA: Cosmo. Sci. Publishers, pp. 175–191.

Cover, T. M. & Thomas, J. A. (1991). *Elements of Information Theory*. New York: Wiley.

Csete, M. & Doyle, J. (2004). Bow ties, metabolism and disease. *Trends. Biotechnol.*, **22**, 446–450.

Doolittle, F. (2000). Uprooting the tree of life. *Set. Am.*, **282**, 90–95.

Dunne, J. A., Williams, R. J., Martinez, N. D., Wood, R. A., & Erwin, D. H. (2008). Compilation and network analyses of Cambrian food webs. *PLoS Biology*, **6** (4), e102, doi:10:1371/journal.pbio.0060102.

Ellis, R. S. (1985). *Entropy, Large Deviations, and Statistical Mechanics*. New York: Springer-Verlag.

Fermi, E. (1956). *Thermodynamics*. New York: Dover.

Fisher, R. A. (2000). *The Genetical Theory of Natural Selection*. London: Oxford University Press.

Forterre, P. (2010). Defining life: the virus viewpoint. *Orig. Life Evol. Biosphere*, **40**, 151–160.

Frank, S. A. (1997). The price equation, Fisher's fundamental theorem, kin selection, and causal analysis. *Evolution*, **51**, 1712–1729.

Freidlin, M. I. & Wentzell, A. D. (1998). *Random Perturbations in Dynamical Systems*, second edn. New York: Springer.

Gell-Mann, M. (1994). *The Quark and the Jaguar: Adventures in the Simple and the Complex*. New York: Freeman.

Gell-Mann, M. & Lloyd, S. (1996). Information measures, effective complexity, and total information. *Complexity*, **2**, 44–52.

Ghosh, K., Dill, K. A., Inamdar, M. M., Seitaridou, E. & Phillips, R. (2006). Teaching the principles of statistical dynamics. *Am. J. Phys.*, **74**, 123–133.

Glasstone, S., Laidler, K. J., & Eyring, H. (1941). *The Theory of Rate Processes*. New York: Mc-Graw Hill.

Gluckheimer, J. & Holmes, P. (1988). Structurally stable heteroclinic cycles. *Math. Proc. Cam. Phil. Soc.*, **103**, 189–192.

Goldenfeld, N. (1992). *Lectures on Phase Transitions and the Renormalization Group*. Boulder, CO: Westview Press.

Goldenfeld, N. & Woese, C. (2011). Life is physics: evolution as a collective phenomenon far from equilibrium. *Ann. Rev. Cond. Matt. Phys.* **2**, 17.1–17.25, doi:10.1146/annurev-conmatphys-062910-140509.

Goldstein, H., Poole, C. P., & Safko, J. L. (2001). *Classical Mechanics*, third edn. New York: Addison Wesley.

Gould, S. J. (1989). *Wonderful Life*. New York: Norton.

Gould, S. J. (2002). *The Structure of Evolutionary Theory*. Cambridge, MA: Harvard University Press.

Gray, H. B. (1994). *Chemical Bonds: an Introduction to Atomic and Molecular Structure.* Sausalito, CA: University Science Press.

Hamilton, W. D. (1964a). The genetical evolution of social behavior. I. *J. Theor. Biol.*, **7**, 1–16.

Hamilton, W. D. (1964b). The genetical evolution of social behavior, II. *J. Theor. Biol.*, **7**, 17–52.

Hamilton, W. D. (1970). Selfish and spiteful behavior in an evolutionary model. *Nature*, **228**, 1218–1220.

Huber, C. & Wächtershäuser, G. (2000). Activated acetic acid by carbon fixation on (Fe, Ni)s under primordial conditions. *Science*, **276**, 245–247.

Hügler, M. & Seivert, S. M. (2011). Beyond the Calvin cycle: autotrophic carbon fixation in the ocean. *Ann. Rev. Marine Sci.*, **3**, 261–289.

Hügler, M., Wirsen, C. O., Fuchs, G., Taylor, C. D., & Sievert, S. M. (2005). Evidence for autotrophic CO_2 fixation via the reductive tricarboxylic acid cycle by members of the ε subdivision of proteobacteria. *J. Bacteriology*, **187**, 3020–3027.

Jaynes, E. T. (1957a). Information theory and statistical mechanics. *Phys. Rev.*, **106**, 620–630. Reprinted in Rosenkrantz (1983).

Jaynes, E. T. (1957b). Information theory and statistical mechanics. II. *Phys. Rev.*, **108**, 171–190. Reprinted in Rosenkrantz (1983).

Jaynes, E. T. (1980). The minimum entropy production principle. *Ann. Rev. Phys. Chem.*, **31**, 579–601. Reprinted in Rosenkrantz (1983).

Kittel, C. & Kroemer, H. (1980). *Thermal Physics*, second edn. New York: Freeman.

Kolmogorov, A. N. (1958). New metric invariant of transitive dynamical systems and endomorphisms of Pebesgue spaces. *Doklady of Russian Academy of Sciences*, **119**, 861–864.

Kolmogorov, A. N. (1965). Three approaches to the definition of the quantity of information. *Problems of Information Transmission*, **1**, 3–11.

Kondepudi, D. & Prigogine, I. (1998). *Modern Thermodynamics: from Heat Engines to Dissipative Structures.* New York: Wiley.

Koonin, E. V. & Martin, W. (2005). On the origin of genomes and cells within inorganic compartments. *Trends Genet.*, **21**, 647–654.

Krakauer, D. C., Collins, J. P., Erwin, D. *et al.* (2011). The challenges and scope of theoretical biology. *J. Theor. Biol.*, **276**, 269–276.

Lengeler, J. W., Drews, G., & Schlegel, H. G. (1999). *Biology of the Prokaryotes.* New York: Blackwell Science.

Lewontin, R. C. (1974). *The Genetic Basis of Evolutionary Change.* New York: Columbia University Press.

Li, M. & Vitányi, P. (2008). *An Introduction to Kolmogorov Complexity and its Applications*, third edn. Heidelberg: Springer.

Ljungdahl, L., Irion, E., & Wood, H. G. (1965). Total sunthesis of acetate from CO_2. I co-methylcobyric acid and co-(methyl)-5-methoxybenzimidazolylcobamide as intermediates with *Clostridium thermoaceticum*. *Biochemistry*, **4**, 2771–2780.

Luisi, P. L. (2006), *The Emergence of Life: from Chemical Origins to Synthetic Biology*. London: Cambridge University Press.

Ma, S.-K. (1976). *Modern Theory of Critical Phenomena*. New York: Perseus.

MacKenzie, R. E. (1984). Biogenesis and interconversion of substituted tetrahydrofolates. In R. L. Blakely & S. J. Benkovic (eds.), *Folates and Pterins*, vol. 1: *Chemistry and Biochemistry of Folates*. New York: John Wiley & Sons, pp. 255–306.

Maden, B. E. H. (2000). Tetrahydrofolate and tetrahydromethanopterin compared: functionally distinct carriers in C_1 metabolism. *Biochem. J.*, **350**, 609–629.

Mahan, G. D. (1990). *Many-particle Physics*, second edn. New York: Plenum.

Martin, W. & Russell, M. J. (2003). On the origin of cells: an hypothesis for the evolutionary transitions from abiotic geochemistry to chemoautotrophic prokaryotes, and from prokaryotes to nucleated cells. *Philos. Trans. Roy. Soc. London*, **358B**, 27–85.

Martin, W. & Russell, M. J. (2006). On the origin of biochemistry at an alkaline hydrothermal vent. *Phil. Trans. Roy. Soc. B*, doi:10.1098/rstb.2006.1881, 1–39.

Martin, W., Baross, J., Kelley, D., & Russell, M. J. (2008). Hydrothermal vents and the origin of life. *Nature Rev. Microbiol.*, **6**, 805–814.

Mézard, M., Parisi, G., Sourias, N., Toulouse, G., & Virasoro, M. (1984). Nature of the spin-glass phase, *Phys. Rev. Lett.*, **52**, 1156–1159.

Morowitz, H. J. (1992). *Beginnings of Cellular Life*. New Haven, CT: Yale University Press.

Morowitz, H. J. & Smith, E. (2007). Energy flow and the organization of life. *Complexity*, **13**, 51–59. SFI preprint # 06-08-029.

Morowitz, H. J., Srinivasan, V., & Smith, E. (2010). Ligand field theory and the origin of life as an emergent feature of the periodic table of elements. *Biol. Bull.*, **219**, 1–6.

Nowak, M. A. & Ohtsuki, H. (2008). Prevolutionary dynamics and the origin of evolution. *Proc. Nat. Acad. Set. USA*, **105**, 14924–14927.

Odling-Smee, F. J., Laland, K. N., & Feldman, M. W. (2003). *Niche Construction: the Neglected Process in Evolution*. Princeton, NJ: Princeton University Press.

Onsager, L. (1931a). Reciprocal relations in irreversible processes. I. *Phys. Rev.*, **37**, 405–426.

Onsager, L. (1931b). Reciprocal relations in irreversible processes. II. *Phys. Rev.*, **38**, 2265–2279.

Peretó, J., López-García, P., & Moreira, D. (2004). Ancestral lipid biosynthesis and early membrane evolution. *Trends Biochem. Sci.*, **29**, 469–477.

Puigbo, P., Wolf, Y., & Koonin, E. (2009). Search for a "Tree of Life" in the thicket of the phylogenetic forest, *Journal of Biology*, **8**, 59, doi:10.1186/jbiol159.

Quastler, H. (1964). *The Emergence of Biological Organization.* New Haven, CT: Yale University Press.

Riehl, W. J., Krapivsky, P. L., Redner, S., & Segrè, D. (2010). Signatures of arithmetic simplicity in metabolic network architecture. *Biol.*, **6** (4), e100725, doi:10.1371/journal.pcbi.1000725.

Rissanen, J. (1989). *Stochastic Complexity in Statistical Inquiry.* Teaneck, NJ: World Scientific.

Rodrigues, J. a. F. M. & Wagner, A. (2009). Evolutionary plasticity and innovations in complex metabolic reaction networks. *PLoS Computational Biology*, **5**, e1000613:1-11.

Rosenkrantz, R. D. (ed.) (1983), *Jaynes, E. T.: Papers on Probability, Statistics and Statistical Physics.* Dordrecht, Holland: D. Reidel.

Russell, M. J. & Hall, A. J. (1997). The emergence of life from iron monosulphide bubbles at a submarine hydrothermal redox and ph front. *J. Geol. Soc. London*, **154**, 377–402.

Russell, M. J. & Hall, A. J. (2006). The onset and early evolution of life. *Geol. Soc. Am. Memoir*, **198**, 1–32.

Russell, M. J. & Martin, W. (2004). The rocky roots of the acetyl-coa pathway. *Trends Biochem. Sci.*, **29**, 358–363.

Schrödinger, E. (1992). *What is Life?: the Physical Aspect of the Living Cell.* New York: Cambridge University Press.

Sherrington, D. (2010). Physics and complexity. *Phil. Trans.R. Soc. A*, **368**, 1175–1189.

Simon, H. A. (1962). The architecture of complexity. *Proc. Am. Phil. Soc.*, **106**, 467–482.

Simon, H. A. (1973). The organization of complex systems. H. H. Pattee (ed.), *Hierarchy Theory: the Challenge of Complex Systems.* New York: George Braziller, pp. 3–27.

Sinai, Y. G. (1959). On the notion of entropy of a dynamical system. *Doklady of Russian Academy of Sciences*, **124**, 768–771.

Smith, E. (1998). Carnot's theorem as Noether's theorem for thermoacoustic engines. *Phys. Rev. E*, **58**, 2818–2832.

Smith, E. (1999). Statistical mechanics of self-driven Carnot cycles. *Phys. Rev. E*, **60**, 3633–3645.

Smith, E. (2003). Self-organization from structural refrigeration. *Phys. Rev. E*, **68**, 046114. SFI preprint # 03-05-032.

Smith, E. (2005). Thermodynamic dual structure of linearly dissipative driven systems. *Phys. Rev. E*, **72**, 36130. SFI preprint # 05-08-033.

Smith, E. (2008a). Thermodynamics of natural selection I: Energy and entropy flows through non-equilibrium ensembles. *J. Theor. Biol.*, **252**, 2, 185–197.

Smith, E. (2008b). Thermodynamics of natural selection II: Chemical Carnot cycles. *J. Theor. Biol.*, **252**, 2, 198–212.

Smith, E. (2011). Large-deviation principles, stochastic effective actions, path entropies, and the structure and meaning of thermodynamic descriptions. *Rep. Prog. Phys.*, **74**, 046601. arxiv.org/submit/199903.

Smith, E. & Morowitz, H. J. (2004). Universality in intermediary metabolism. *Proc. Nat. Acad. Set. USA*, **101**, 13168–13173. SFI preprint # 04-07-024.

Smith, E. & Morowitz, H. J. (2010). The autotrophic origins paradigm and small-molecule organocatalysis. *Orig. Life Evol. Biosphere*, **40**, 397–402.

Srinivasan, V. & Morowitz, H. J. (2009a). Analysis of the intermediary metabolism of a reductive chemoautotroph. *Biol. Bulletin*, **217**, 222–232.

Srinivasan, V. & Morowitz, H. J. (2009b). The canonical network of autotrophic intermediary metabolism: minimal metabolome of a reductive chemoautotroph. *Biol. Bulletin*, **216**, 126–130.

Stock, G., Chosh, K., & Dill, K. A. (2009). Maximum caliber: a variational approach applied to two-state dynamics. *J. Chem. Phys.*, **128**, 194192:1–12.

Stryer, L. (1981). *Biochemistry*, second edn. San Francisco, CA: Freeman.

Touchette, H. (2009). The large deviation approach to statistical mechanics. *Phys. Rep.*, **478**, 1–69. arxiv:0804.0327.

Vetsigian, K., Woese, C., & Goldenfeld, N. (2006). Collective evolution and the genetic code. *Proc. Nat. Acad. Set. USA*, **103**, 10696–10701.

Wächtershäuser, G. (1988). Before enzymes and templates: a theory of surface metabolism. *Microbiol. Rev.*, **52**, 452–484.

Wächtershäuser, G. (1990). Evolution of the first metabolic cycles. *Proc. Nat. Acad. Set. USA*, **87**, 200–204.

Weinberg, S. (1995). *The Quantum Theory of Fields*, Vol. I: *Foundations*. New York: Cambridge.

Weinberg, S. (1996). *The Quantum Theory of Fields*, Vol. II: *Modern applications*. New York: Cambridge.

Wilson, K. G. & Kogut, J. (1974). The renormalization group and the ε expansion. *Phys. Rep., Phys. Lett.*, **12C**, 75–200.

Woese, C. R. (1987). Bacterial evolution. *Microbiol. Rev.*, **51**, 221–271.

Woese, C. R. (2000). Interpreting the universal phylogenetic tree. *Proc. Nat. Acad. Set. USA*, **97**, 8392–8396.

Wu, D., Ghosh, K., Inamdar, M. *et al.* (2009). Trajectory approach to two-state kinetics of single particles on sculpted energy landscapes. *Phys. Rev. Lett.*, **103**, 050603: 1–4.

10   The inferential evolution of biological complexity: forgetting nature by learning to nurture

David Krakauer

Even the recognition of an individual whom we see every day is only possible as the result of an abstract idea of him formed by generalization from his appearances in the past.

James G. Frazer

I don't paint things. I only paint the difference between things.

Henri Matisse

Mathematics is the art of giving the same name to different things.

Henri Poincaré

10.1 MONDRIAN AND DIRAC ON MINIMALISM

The Dutch painter Piet Mondrian is best known for strikingly simple canvases populated by vertical and horizontal planes rendered in primary colors. Mondrian's paintings from the 1930s: *Composition with Yellow Patch, Composition No. 8*, and *Vertical Composition with Blue and White*, are a far cry from his earliest canvases, *Landscape with Ditch, Mill in Sunlight*, and *Red Tree*. The early canvases feature colorful and vigorous representations of scenes, and are indebted to the post expressionists, Suerat and Van Gogh. These in time give way to the later compositions influenced by the analytical abstractions of the cubists, Picasso and Braque. The names of the later paintings reveal a parallel trend towards descriptive simplicity. In his correspondence, Mondrian provides a clue to his thought leading to this change of style: "The principle of this art . . . is not a negation of matter, but

Complexity and the Arrow of Time, ed. Charles H. Lineweaver, Paul C. W. Davies and Michael Ruse. Published by Cambridge University Press.
© Cambridge University Press 2013.

a great love of matter, whereby it is seen in the highest, most intense manner possible, and depicted in the artistic creation." (1912) and "We arrive at a portrayal of other things, such as the laws governing matter. These are the great generalities which do not change." (1913) (Faerna, 1997). Over the course of the cultural evolution of Mondrian's creative life, we observe a systematic trend away from detail and realism and towards minimalism and abstraction. Painterly subjects easily recognized in his early canvases become a kind of affine symbolism in the later paintings. The evolution of Mondrian's style is towards abstract ideas that preserve only the barest, nominal distinctions among his subjects.

Contemporary with Mondrian, the English physicist Paul Dirac sought to uncover fundamental correspondences and connections between the newly emerging field of quantum mechanics and the recently confirmed predictions of general relativity (Farmelo, 2011). For his work in quantum mechanics, Dirac shared the Nobel Prize in 1933. In the presentation speech by Professor H. Pleijel, Chairman of the Nobel Committee for Physics of the Royal Swedish Academy of Sciences, Dirac's contributions were described in these terms: "Dirac has set up a wave mechanics which starts from the most general conditions. From the start he put forward the requirement that the postulate of the relativity theory be fulfilled. Viewed from this general formulation of the problem it appeared that the self-rotation of the electron which had previously come into the theory as an hypothesis stipulated by experimental facts, now appeared as a result of the general theory of Dirac." As with Mondrian, Dirac favored extreme forms of abstraction. Whereas his contemporaries, Schrödinger and Heisenberg, preserved in their representations of the physical world some semblance of physical reasoning building on the intuitive foundations of classical mechanics, Dirac was willing to forfeit all forms of material correspondence in favor of abstract relationships among mathematical structures (Mehra & Rechenberg, 2000). Dirac explained his preference for abstraction, "I learnt to distrust all physical concepts as the basis for a theory. Instead one should put one's trust in a mathematical scheme, even if the scheme does not appear at first sight to be connected with physics. One should concentrate on getting interesting mathematics."

Both Mondrian and Dirac thought, in the words of da Vinci, "simplicity is the supreme sophistication". Both Mondrian and Dirac moved closer to reality by moving away from its direct representation. The evolution of the art of Mondrian and the science of Dirac does not demonstrate an increase in complexity, but a striving toward its opposite. For both scientist and artist, simplicity achieved greater generality and a more efficient encoding of the fundamental nature of reality.

This is an essay on biological complexity that I have chosen to introduce with a detour into cultural evolution, or devolution. My purpose is to establish a counter-intuitive realization – that sophistication or functional efficacy are not equivalent to complexity. We tend to reflect on biology from the vantage point of an anthropomorphic assumption – that a human nature, a complex inferential mechanism, is the most likely outcome of a cumulative, evolutionary process. Cumulative evolutionary processes can also yield simplicity. Mondrian represents a stylistic point on a long lineage passing through Lascaux, Corinthian curlicues, the golden Mozaics of the Byzantine period, the mythic landscapes of Botticelli and the splendors of recumbent Odalisques by Ingres. Dirac similarly descends from the mathematical mysticism of Kepler, the rigorous iconoclasm of Galileo, the universal genius of Newton, and the mind-bending intuition of Einstein. But rather than resembling the scholarly equivalent of a barnacle-encrusted hull of an ancient caravel, built up layer upon layer by sedimentation, Mondrian and Dirac created a parsimonious minimalism faithful to history and brutally minimal.

There is no self-evident reason why cultural evolution should increase in complexity just because it builds on a long history. And the same is true for biology, which, through mutation, drift and natural selection, adopts and disposes of the past in such a way as to promote an approximate best fit to the present. If anything, there is in simplicity once arrived at, an efficiency often lacking in more complex representations, and for this reason we might expect simplicity to be favored over complexity (Gell-Mann, 1995).

On the other hand, life does seem in some fundamental way to be different from art. Life on Earth started in a rather minimal protocellular state, and has evolved to include a profusion of multicellular species with each individual constituted by a dense network of ecological interactions (Smith & Szathmary, 1999; Davies, 2006). If we operationalize complexity informally, in terms of the diversity of parts and interactions, then through time there are evidently lineages that have increased in such quantities (McShea, 1991). But there are others that have experienced a corresponding reduction – demonstrated an inclination toward simplicity (Gould, 2002). An evolutionary theory for complexity should be able to explain those circumstances when accretion takes place, and when it does not. A powerful theory might explain how it is possible for simple organisms to accomplish the same functions as complex organisms while escaping replacement and extinction. An evolutionary theory of complexity must, in other words, account for both simplicity and complexity, and provide conditions when each of these outcomes is favored – appreciating when simplicity is the most efficient solution to a complex problem – when the solutions of a Mondrian or a Dirac hold sway over those of a Delacroix or a Maxwell.

10.2 THE NATURE OF EVOLUTIONARY EXPLANATION

Ruse, in a recent synoptic essay on the history of evolutionary thought (Ruse & Travis, 2009), describes evolutionary explanations as all of those arguments that adduce natural means of accounting for biological diversity. Evolutionary arguments seek to: (1) describe the features of organisms in terms of demonstrable functions/adaptations, and (2) establish continuity among organisms by tabulating taxonomically shared features. In practice, this consists of mapping horizontally

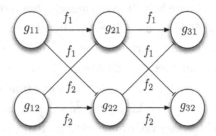

FIGURE 10.1 Evolution as a directed, acyclic graph. Genomes in successive generations encode functions that enhance their own proliferation at the expense of competitors. Evolution by natural selection is therefore a heterogeneous flow of genotypes forward in time. The f_i encode the replication rates of each genome, or their functional contributions or weights in the directed graph, both positive (lines ending with an arrow) to their descendants in the future and negative (lines ending with bars) to their competitors.

(analogically) features of a different species onto mechanisms in support of various adaptive functions (hearts onto pumps), and mapping vertically (homologically) features of an organism onto those of putative common ancestors (wings onto arms). A third form of argument seeks to find evidence of trends along the vertical or temporal axis corresponding to more and more elaborate mechanisms. These have been described as trends towards increasing complexity (and in social contexts, progress (Ruse, 2009)).

Darwin's insight into selection was to provide a mechanism for changes in character traits that can accumulate via differential survival of organisms across generations (time). This is illustrated in a very simple case in Fig. 10.1 as a directed graph, where features j of generation i are indicated as g_{ij} and functions as f_j. This network represents an elementary Darwinian process by virtue of assigning negative (inhibitory or competitive) values to the diagonal f_j terms. Hence the Darwinian framework describes the changing features of populations of organisms over time, and explains these in terms of both inherited features of ancestors and the differential success of features across a range of environments. There is obviously nothing in this picture that suggests a trend in the values of g_{ij}.

For a population of N different organisms, this scheme can be translated directly into evolutionary dynamics in the form of the replicator equation (RE), which in continuous time drops the generational index, preserving only the function, and imposes a constraint on total population size through density dependent regulation (negative self-interaction).

$$\dot{g}_i = g_i \left(f_i - \sum_j f_j g_j \right).$$
(10.1)

By introducing mutation rates q_{ij} where type $j \to i$ we allow for one to many mappings across generations, and drop the unrealistic requirement that all diversity be present as an initial condition. This generates an expanded system of equations described as the replicator–mutator equation, or quasi-species equation (Nowak, 2006):

$$\dot{g}_i = \sum_{j=1}^{N} g_j f_j q_{ij} - g_i \sum_{i}^{N} f_i g_i,$$
(10.2)

where $q_{ij} = p^{H_{ij}}(1 - p)^{L - H_{ij}}$, the value H_{ij} in the Hamming distance between genomes i and j, and L is the length of the genome composed of the four canonical nucleotides.

The point of this little mathematical interlude is to demonstrate that there is little obvious in the way of a space-time trend in Darwinian evolution contemplated from the perspective of equations of motion, toward neither complexity or simplicity. All changes derive from properties of fitness and mutation (and drift with finite populations). All that we can explain is temporal variation in the composition of a population of density-limited replicators. According to one popular short text in population genetics, this is all that we should expect, as "Evolution is the change in frequencies of genotypes through time, perhaps due to their differences in fitness" (Gillespie, 2004).

I shall argue in this chapter that there is something rather fundamental lacking in this kind of definition. Whereas the definition recounts a truism, it manages to side-step the idea that frequencies often change in time in order to more effectively predict the state of the environment. This ability to predict is key to understanding long-term trends, and why we might observe changes in genome size or in the diversity of cell types. In other words, prediction is implicit in the fitness term, which in these formalisms is encoded through a bias in the rates of genome replication.

Evolutionary biology as a discipline has been very successful in addressing questions of changing frequencies, and consequently developed a mental moat around this core feature, keeping out complementary ideas that are equally fundamental (see e.g. Carroll, 2001). Attempts to address ideas of function are often relegated to the atemporal, physiological, and behavioral sciences such as molecular and cognitive biology, which emphasize the mechanistic underpinnings of functional traits. A possible reconciliation is to recognize that these fields explore analogous phenomena at different scales of time. The division of cells within a body in response to growth signals, and the formation of connections in the brain in response to sensory feedback, could also be defined in terms of tracking changing frequencies – of cell types and connections – in response to somatic-selection signals. Fundamental to these fields are the homeostatic and regulatory implications of changes in frequency. Evolution in a larger sense might better be described as the changing frequency of alternative environment-predictors. Predictions that meet with failure fail to survive.

In the following sections I will devote a few pages to exploring the implications of simple Darwinian dynamics. (A more detailed account is in Krakauer, 2011.) These dynamics reveal significant obstacles to the evolution of complex forms of life. My working model is that evolution, like a gifted artist, generates novel solutions by working within and overcoming fundamental constraints. I will very briefly describe the general way in which these constraints are overcome through the evolution of basic cognitive capabilities: mechanisms that augment the predictive capacity of genomes alone. For our purposes, complexity will be a measure of inferential or predictive capacity: the ability for a population to store adaptive responses to a variety of states of the environment.

10.3 SELECTION, INFORMATION, AND DEATH

Let's address the question of evolutionary trends in general by considering the one trend that is implicit in all evolutionary dynamics. Darwin's theory of natural selection is a theory that introduces order into an individual genome by disordering a population. Fixation of a selected variant is achieved through the mortality of alternative genotypes.

Consider a one-dimensional, discrete space of resources – food, partners, shelter. At each position there is a function of that resource f_i. A given genotype encodes an organism's prediction of the state of the environment. In Fig. 10.1 we described these as weights mapping genotypes onto themselves and inhibiting competitors. Populations distribute themselves over this functional resource space $J = \{1, \ldots, n\}$, generating in a large enough population a distribution over the entire functional space, g_i. So g_i is an investment in the function f_i. Selection in a population of fixed size is described by the RE,

$$
\dot{g}_i = g_i \left(f_i - \sum_j f_j g_j \right),
$$

$$
s_N = \left\{ \mathbf{g} \in \mathbb{R}^N : g_i \geq 0 \forall_i \in J, \sum_i g_i = 1 \right\}.
$$

We define the amount of information gathered (IG) by selection at time t as the difference between the maximum possible certainty about the choice of function and the observed information. This can be written in terms of standard Shannon entropies:

$$
IG(t) = \sum_i^N g_i(t) \log g_i(t) - \log(1/N),
$$

where the function $IG(t)$ is the information gained at time t. The solution to the RE is unique and depends only on the relative values of f_i, such that

$$g_i = 1, \quad i = \max_i(f_i),$$
$$g_j = 0, \quad \forall_j \in n\backslash i.$$

The amount of information that is acquired for performing a preferred function is directly proportional to the loss of life. The ability to find an adaptive peak requires that all resources are channeled into a single genotype. This can be stated slightly more formally,

$$IG(t \to \infty) = \log(N). \tag{10.3}$$

Around N variants must be lost to obtain $\log(N)$ quantity of information. It follows that the greater the number of possible states of the environment, the more variant genotypes there must be in order to match them. Prediction of the environment requires memory of the possible states of the environment in the population.

10.4 EVOLUTION AND COMPLEXITY

Continued evolution implies that the environment is constantly changing. Consider a case where the state of the environment is sampled from some known probability distribution. A source of free-energy f_i on a one-dimensional lattice of dimension N is now a continuous random variable, where $\Pr[0 \le f_i \le F] = \int_0^F \theta(f_i)df_i$. At an interval of several generations the lattice is periodically reconfigured, sampling from the distribution $\theta(f)$.

We previously established that evolutionary dynamics can maximize the information content of the genome. Hence selection will seek to maximize the entropy $H(L)$ subject to constraints on the value of L. We also know that at equilibrium, for $\log(N) = L$, there is little or no information in the environment that is not present in the genome. Over evolutionary time the genome is itself evolving and there will be some probability distribution over the 4^L states of the genome, call this $\hat{L}$. Through selection the genome records the lattice position that is most common for a given environment. This implies that selection seeks to minimize equivocation or conditional entropy $H(\hat{L}|\theta)$, capturing all environmental information that is not present in the genome. We define the complexity of our system as the maximum of the difference between the information in the genome and the equivocation,

$$C = H(\hat{L}) - H(\hat{L}|\theta) \tag{10.4}$$

$$= H(\hat{L}) + H(\theta) - H(\hat{L}, \theta). \tag{10.5}$$

This quantity is known as mutual information and measures the maximum amount of information that selection can transfer into genomes from the environment. Adami first proposed this quantity as a natural measure of evolutionary complexity (Adami, Ofria, & Collier, 2000), focusing on the universality of the genetic code as a natural basis for comparison. We note that there can be no information in the genome that is not already present in the environment. Any state of the genome that is not under selection will simply drift towards a maximum entropy state. We find that as the information content of the environment increases, so does this measure of complexity. We also understand that whereas $H(\hat{L})$ is bounded by free-energy constraints on L, we are not aware of similar bounds on $H(\theta)$. The joint entropy $H(\hat{L}, \theta)$ ensures that organismal complexity cannot grow beyond the limits imposed by the maximum L, as for any preferred state of an environment where $\log (N) > L$ there will be a maximum of ignorance over which state selection prefers. Similarly, increasing the rate of error (e.g. mutation) will increase $H(\hat{L}, \theta)$ by randomly assigning genotypes to states of the environment. This measure of complexity is ensemble-based and requires that we take measurements over many genomes in a variety of environments. In other words, evolutionary complexity, rather like evolution itself, is not a property of an individual but a population. Any one lineage will evolve towards a single best strain, and the information about the probability distribution over environmental states will be lost.

Mutual information is valuable as a measure of evolutionary change in a similar way to the value of an energy function in a Hamiltonian formulation of a conserved system, or a Lyapunov function in a dynamical system. Each of these measures provides a function that the system can be observed to maximize or minimize. By expressing our dynamical system in terms of an information theoretic function, we have not predicted anything beyond the dynamical formulation, but we have shown what the dynamics are yielding at a global level. The informational quantity gives us a new insight into the "semantics" or function of the evolutionary process, and provides a rigorous measure of the effectiveness of these dynamics.

Since we have observed that mutual information can be observed through evolutionary dynamics, it is natural to ask why lineages are not all maximally complex with regard to their mutual information with the environment. Why do some lineages become "living fossils" and preserve a fixed phenotype for very long spans of time without unambiguous evolutionary trends? And why, under certain conditions, does complexity reduce through time, with some organisms evolving toward extreme simplicity?

I consider all three possibilities and show that they are perfectly compatible with the same complexity measure.

(1) *The preservation of simplicity*   The most obvious explanation for simplicity is that the discernible environmental variability (the number of states of the environment) remains low. Not all organisms experience the same environment such that even "simple" – low $H(\hat{L})$ – lineages could be at the maximum of complexity given local environmental variability. There is simply no further information available to be extracted.

(2) *The constancy of complexity*   There is no need to increase organismal information content when equivocation has been minimized and regulatory information maximized. If an environment is changing for long stretches of time within a fixed envelope of variability, then the same should be true for the organism.

(3) *The reduction of complexity*   There are two scenarios that should lead to a reduction of complexity. The first is a reduction in the possible

states of the environment. If N is reduced, then any marginal cost associated with increasing L should lead to a reduction in $H(\hat{L})$. This is beautifully exemplified in parasite evolution, where the free energy probabilities of a "lattice site" (trait value) become zero or one. Hence if an enzyme required to complete the virus life cycle is predictably encoded by the host, or more generally if some feature of the virus becomes redundant, there is no need for the virus to encode this conditional trait. This will lead to a reduction in $H(\hat{L})$. The second scenario corresponds to an increase in noise or a reduction in free energy available for error correction. This will tend to increase $H(\hat{L}, \theta)$ and reduce organismal complexity through increased dissipation of information.

10.4.1 Complexity and simplicity

We now have a fairly intuitive framework for exploring the evolution of complexity and simplicity. When the states of the environment are sufficiently small, there is a corresponding minimal need to encode or "remember" alternative genetic configurations. As these states grow, there is a benefit in remaining capable of tracking these states. But the number of states of the environment is not sufficient to explain simplicity and complexity. We also need to attend to the predictability of these states. Completely predictable states of the environment do not require conditional responses, only a fixed feature of the organism. Completely random states can not be predicted by definition, and are better off ignored. So increasing the number of regular but not fixed features of the environment is predicted to lead to an increase in the number of adaptive states of the organism. Contrariwise, reducing the number of regular states, or moving towards regimes of perfect predictability or randomness, lead to a reduction in the number of adaptive states of the organism.

But there has to be more to the story. This approach might help to explain why genomes and their regulation become more complex, but it does not explain why life did not stop at the single cell. That is why multicellular organisms evolved to solve the same problem over shorter timescales by evolving nervous systems and a variety of

mechanisms of learning and cultural transmission. The key to understanding this "major transition" is a fundamental limit to the size
of L generated by the timescale of information flow through natural
selection (see Krakauer, 2011 for a fuller discussion of this limit in
relation to the idea of Maxwell's demon). The mutual information
of evolution is measured in bits per generation, which are not much
good for an organism that seeks to learn from the environment within
a single lifetime.

10.4.2 The genetic bottleneck

Define the probabilities of mutating among genomes by considering a
simple bit-wise operation, in which the probability of any one element
of the genome of length L mutating over the course of one generation
is given by a probability p.

$$q_{ij} = p^{H_{ij}}(1 - p)^{L - H_{ij}}. \tag{10.6}$$

The value H_{ij} is the Hamming distance between genomes i and j,
and the binomial distribution of mutation rates q_{ij} arises because a
transition between any two genomes becomes exponentially more
rare the further they are apart in sequence space.

It is straightforward to deduce the threshold genome length
below which information can be preserved – the mutation rate below
which selection can favor one genotype over all others (Eigen, 2000):

$$L < \frac{1}{p}.$$

The only way new information can be acquired is through mutation and fixation, but the greater the rate of mutation p, the lower the
total storage capacity of the genome L. This is an inescapable constraint operating on a genome, and requires mechanisms of selection
or learning other than natural selection in order to be overcome.

10.4.3 THE COGNITIVE CAPABILITY

How might an organism that is driven to acquire and store more
information to survive discover a means of overcoming the genetic

constraints imposed by mutation and selection? I suggest that this can be achieved through the evolution of mechanisms that possess three key properties. These properties are observed both at the level of single cells, and in large tissue systems such as brains.

(1) Self-similar structure generated by minimal (or short) genetic regulatory networks.
(2) Modification of these structures within a single generation using suitable environmental cues.
(3) A property of arbitrary assignment of modifications to structures.

These conditions are required because we desire a surrogate for L which does not scale with L. We require that multiple fixation events can take place within a single generation. And we require that adaptive information can be stored in some form that is independent of the source of the information – the information has the linguistic property of arbitrariness.

Self-similarity
Self-similarity is widespread in adaptive systems and contributes to many of the fundamental scaling laws in biology. Self-similarity is found from the level of DNA supercoiling, up through to the branching structure of the neurons in the brain. Self-similarity can be as simple as the serial repetition of an element, through to the generation of complex, nested shapes. One of the key features of self-similarity is that arbitrarily large, hierarchical structures can be generated by iterating a simple developmental rule. This is very nicely illustrated through the method of fractal compression (Barnsley & Hurd, 1993). Regular, lossy compression describes methods of data encoding in which data is represented up to some arbitrary accuracy by a significantly reduced data set. Compression make use of redundancies in source data in order to achieve efficiencies in a target data set. Fractal compression exploits the hierarchical structure of source data (repeating features at different scales in a visual scene, for example), and represents the data through an iterated function system. This method of compression for visual scenes has a convenient property

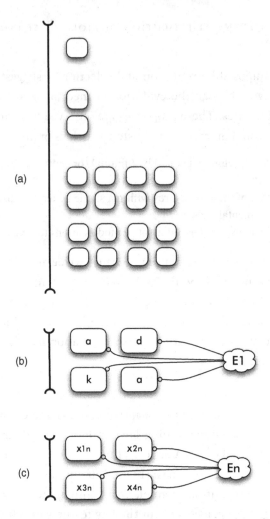

FIGURE 10.2 The three cognitive requirements. (a) Infinite structure from finite information, realized through self-similar patterning. By repeating the expression of a gene, we can build increasingly large structures without having to build larger genomes. This is indicated here through a simple duplication of a pattern of basic elements (achieved for example through multiple translation events of a single, fixed-length RNA template). (b) The ability to imprint environmental states. Each of these environmental states is indexed (E_i) by conditionally activating a variety of elements within a repeated structure (indicated with lower case letters). Each element has a preference for a given state (a, d, k) in a given environment. (c) The semiotic, or arbitrary property of the imprint to structure mapping, allowing any element with a unique address (x_i) to encode a given environmental state (n) for future readout. This kind of imprinting allows for an effectively unlimited amount of information to be stored on the self-similar "pages". This is characteristic, for example, of chromatin modification where histone tails are modified by a range of enzymes sensitive to environmental conditions.

that it can generate resolution independent target images from a finite source image.

The developmental encoding of the lungs, circulatory system, and brain all make use of the small footprint property of fractals. Making a larger lung does not require a larger genome, simply more steps of cell division and morphogenesis. And making a larger brain, with a corresponding increase in memory and processing capacity, does not require a larger genome. And having a larger genome would not help, as single cells lack the ability to identify states of the environment at large space and timescales. Sensory organs exploit cooperation among many cells to detect large-scale features of the environment.

Imprinting

With a large self-similar structure at our disposal, we need a means of systematically modifying it, or making it sensitive to environmental regularities. The central dogma of molecular biology largely rules this out (at least empirically) for DNA and RNA. But there is another path. At the molecular level DNA is always associated with histone proteins forming the chromatin complex. Histone proteins possess long tails that can be biochemically modified. Changing the chemical environment of the cell leads to changes in the chemical composition of chromatin. Thereby DNA can be secondarily tagged along a repeating structure that is potentially as information-rich as the genome proper, and possesses the benefit of being directly modifiable within a single cell or organismal generation.

At a cellular level, many cells possess excitable membranes that respond to changes in their ionic environment. These changes can lead to regular patterns of electrical activity, such as in action potentials. Nervous systems are large self-similar structures that are capable of recording, through chemical and electrical modifications, salient features of their environments.

Arbitrary tagging

It is critical that the structure not impose excessive constraints on the form of the imprinted modifications. Hence with chromatin there are multiple positions on any given histone protein, and large numbers of histones that can be modified in many different ways. The aggregate modification and their context bias gene expression. For neurons in the brain the same is true – electrical activity among populations of cells can be used to represent many environmental features encoded as spike trains. In both cases the same molecules or tissues can be "written", and subsequently "read" in a very large combination of different ways.

Using cognitive traits to learn

A cognitive mechanism lends itself to exactly the same complexity definition that we used for evolutionary dynamics. Operant conditioning is one way in which animals have been observed to learn to perform a rewarded behavior. Rather than a set of alternative genomes, consider a set of alternative behaviors X_i each associated with a probability x_i. For each behavior there is an environmental reward r_i. A feature of the X vector is that the dimension can be independent of the dimensions of L: many neurons can be made from the same developmental plan, all we need do is increase the number of rounds of cell division to obtain a larger number of memory states. Larger brains do not require larger genomes to be encoded. The incremental change in the probability of any action, with a learning rate parameter α, is often written as (see Zhang, 2009 for a full derivation for operant conditioning and replicator dynamics):

$$\delta x_k = \alpha r_i (e_{ik} - x_k), \tag{10.7}$$

where the e_{ij} is the Kronecker δ function,

$$e_{ik} = \begin{cases} 1, & \text{if } k = i, \\ 0, & \text{if } k \neq i. \end{cases}$$

This gives the updated probability of behavior assuming that a single action was performed. This is analogous to an increase in gene frequency arising through selection. The rate of update is no longer measured in units of generations but units of temporal action. The average change of behavior takes into account the probability of an action x_i,

$$\Delta x_k = \sum_i x_i \delta x_k. \tag{10.8}$$

Combining these two equations, we deduce that

$$\Delta x_k = \sum_i x_i [\alpha r_i (e_{ik} - x_k)] \tag{10.9}$$

$$= \alpha \sum_i x_i r_i e_{ik} - \alpha x_k \sum_i x_i r_i \tag{10.10}$$

$$= \alpha x_k r_k - \alpha x_k \sum_i x_i r_i \tag{10.11}$$

The term $\sum_i x_i r_i$ is simply the mean reward, $\langle r \rangle$, and the change in behavior in continuous time is given by

$$\dot{x}_k = x_k (r_k - \langle r \rangle). \tag{10.12}$$

Behaviors that receive rewards that are greater than the average reward will be reinforced, and those obtaining a lower reward will be extinguished. As with the replicator equation for genetic replication and selection, this learning dynamic lives in the real valued unit simplex. And in analogy with natural selection, learning will tend to maximize the amount of information that an individual extracts from the environment. The best behavior will fix in the population excluding all others until the environment changes.

If we define the distribution over the number of states B of the cognitive systems as $\hat{B}$ and allow for a distribution θ_n of environment states detectable by a neural system, we could, in analogy with evolutionary complexity, define neural complexity as

$$C_n = H(\hat{B}) - H(\hat{B}|\theta_n) \tag{10.13}$$

$$= H(\hat{B}) + H(\theta_n) - H(\hat{B}, \theta_n); \tag{10.14}$$

the same logic for complexity and simplicity applied to the genome. Regularity without determinism will tend to increase C_n, whereas randomness or full determinism will tend to reduce C_n.

10.4.4 COGNITIVE COMPLEXITY

When we combine the empirical observation that organisms seek to record the maximum of adaptive information within their genomes, while evolving mechanisms to overcome the limitations of genomes and selection, we obtain important insights into the evolution of biological complexity. Biological complexity is hypothesized to be the competitive evolution of mechanisms that overcome genomic bottlenecks by means of scaleable structures capable of representing a full spectrum of environmental regularities using arbitrary imprints. In other words, biological complexity is thought of as a series of cognitive innovations, all aimed at supplementing or overcoming the adaptive limitations of genomes, allowing for more efficient representations and predictions of environmental regularities inducing fitness consequences.

The complexity of the organism has gone from C to $C + C_n$. Information from the environment that changes only infrequently (relative to an organismal generation) is captured in the measure C, whereas information that changes quickly will be captured in the measure C_n. This distinction is exactly what Chomsky had in mind when he spoke of principles and parameters of natural language. The "universal" principles contribute to C whereas the culture-dependent parameters are recorded in C_n.

10.4.5 FORGETTING NATURE AND LEARNING TO NURTURE

The great debate over nature versus nurture is an argument resting on a false premise. The premise asserts that some traits of the organism are fixed whereas others are variable. The premise is false because nothing about a species remains fixed over suitably long periods of time. As Keynes (1923) put it, "The long run is a misleading guide to current affairs. In the long run we are all dead". We might, however, be interested in establishing what can be learnt quickly by individuals versus what can only be learnt slowly by populations (evolution by natural selection). After all, over time some lineages have supplemented natural selection with mechanisms that perform the same function at a faster rate and at a higher capacity. I have called these mechanisms cognitive, and they consist of mechanisms that can imprint information on scalable structures that need not increase indefinitely with genome size.

The cognitive ability to learn removes a significant burden from genomes and selection, allowing that nature (slow replicative timescales) can be forgotten when the same information can be nurtured through learning (fast ontogenetic timescales). So returning to the epigraphs at the start of this chapter, when Henri Poincaré wrote "Mathematics is the art of giving the same name to different things", he captured the essence of the idea in this chapter, that evolution and learning are both inferential dynamics that build representations in order to predict their environments. And when Henri Matisse wrote "I don't paint things. I only paint the difference between things", he was suggesting that we only need to encode the informative bits, not the redundancies. Cognitive systems evolve to encode regularities that can not be captured efficiently by a genome. And both genomes and brains use these informative differences from the past to make informed predictions about the future, as – James G. Frazer put it: "Even the recognition of an individual whom we see every day is only possible as the result of an abstract idea of him formed by generalization from his appearances in the past".

It is of outstanding interest to establish whether evolutionary processes, grounded in purely material mechanisms, are expected to generate complex forms of life. My preferred definition of complexity is based on the ability to encode and thereby predict the environment – cognitive capacity. This more or less corresponds to the common idea, at least as directed towards our own species, expressed here by Thoreau: "I know of no more encouraging fact than the unquestionable ability of man to elevate his life by conscious endeavor." I have attempted to show that there are necessary conditions for an increase in cognitive capability. These are when regularities in the environment present themselves at a rate that exceeds the ability of natural selection to record an adaptive response in the genome. Under these conditions, epigenomes and nervous systems can grow to capture information, and will often do so with an attendant loss of information in the genome. When environments change slowly, cognition offers little advantage and genomes should be sufficient to encode adaptive responses. Complexity therefore increases on a "need to know" basis, and does not follow an inexorable trend but a wavering pathway governed by the rate of information production by the environment, where the environment includes organisms and their products. Our measure of complexity should be sufficiently robust to explain how improving adaptive ability can lead to both increases and reductions in complexity. Mondrian is not the lesser artist, and Dirac is not the lesser scientist than their more "realistic" and "detailed" contemporaries. Both appropriate representational styles and solutions to compressible observations.

In the same way that the evolution of wings and flight are compatible with evolution by natural selection but not the fate of all lineages, complex cognitive capacities might evolve where and when the conditions of a species support high rates of information flow between organism and environment. Much as we might like to find flight when it is absent, as in Cymbeline's lament, "O! for a horse with wings!", we perfectly understand that there are costs and benefits to all traits, and that the possibility of a Pegasus is not the same thing as its occurrence in nature. Our future understanding of the evolution of complexity will consist of compiling systematically an inventory of conditions and measurements where complexity is likely to increase or decrease – probabilistically, not inevitably – and searching for these conditions on Earth and, optimistically, beyond our biosphere.

REFERENCES

Adami, C., Ofria C., & Collier, T. (2000). Evolution of biological complexity. *Proc. Nat. Acad. Sci.*, **6**, 363–376.

Barnsley, M. & Hurd, L. (1993). *Fractal Image Compression*. A. K. Boca Raton, FL: Peters/CRC Press.

Carroll, S. (2001). Chance and necessity: the evolution of morphological complexity and diversity. *Nature*, **409**, 1102–1109.

Davies, P. (2006). *The Origin of Life*. London: Penguin.

Eigen, M. (2000). Natural selection: a phase transition? *Biophysical Chemistry*, **85**, 101–123.

Faerna, J. M. (1997). *Mondrian Harry N*. New York: Abrams.

Farmelo, G. (2011). *The Strangest Man. The Hidden Life of Paul Dirac, Mystic of the Atom*. New York: Basic Books (AZ).

Gell-Mann, M. (1995). *The Quark and the Jaguar. Adventures in the Simple and the Complex*. New York: St. Martin's Griffin.

Gillespie, J. (2004). *Population Genetics: a Concise Guide*. Baltimore: JHU Press.

Gould, S. (2002). *The Structure of Evolutionary Theory*. Cambridge, MA: Harvard University Press.

Keynes, J. M. (1923). *A Tract on Monetary Reform*. Amherst, NY: Reprinted by Prometheus Books (April 2000).

Krakauer, D. C. (2011). Darwinian demons, evolutionary complexity, and information maximization. *Chaos*, **41**, 037110–037110-12.

McShea, D. W. (1991). Complexity and evolution: what everybody knows. *Biology and Philosophy*, **6**, 303–324.

Mehra, J. & Rechenberg, H. (2000). *The Historical Development of Quantum Theory*. New York: Springer.

Nowak, M. A. (2006). *Evolutionary Dynamics: Exploring the Equations of Life*. Cambridge, MA: Belknap Press.

Ruse, M. (2009). *Monad to Man. The Concept of Progress in Evolutionary Biology*. Cambridge, MA: Harvard University Press.

Ruse, M. & Travis, J. (2009). *Evolution. The First Four Billion Years*. Cambridge, MA: Belknap Press.

Smith, J. M. & Szathmary, E. (1999). *The Origins of Life. From the Birth of Life to the Origin of Language*. New York: Oxford University Press.

Zhang, J. (2009). Adaptive learning via selectionism and Bayesianism, Part I: Connection between the two. *Neural Networks*, **22**(3) (April): 220228. doi:10.1016/j.neunet.2009.03.018.

11 Information width: a way for the second law to increase complexity

David H. Wolpert

11.1 RISING COMPLEXITY, NATURAL SELECTION, AND THERMODYNAMICS

It seems that many systems both increase their complexity if initialized in a low complexity state, and then reliably and robustly maintain high (but finite) complexity once it is attained. Some of the most prominent examples are the many biological systems undergoing natural selection that seem to start with low complexity and then increase their complexity (Krakauer, 2011; Carroll, 2001; McShea, 1991; Smith, 1970). Such systems are typically modeled as localized individuals that reproduce in an error-prone process, with their offspring weeded out in competitions with other individuals that select for higher complexity. In this natural selection process the individuals in a line of biological descent increase their complexity in time.

Some have argued from these examples that natural selection is a necessary condition for complexity to increase. The idea is that for a particular lineage to have a large fitness advantage over its competitors, it must become increasingly "complex". However, we should not confuse the properties of an example of a phenomenon with the phenomenon itself: complexity increase is not synonymous with adaptionist natural selection. Indeed, one can engineer by hand models of systems undergoing natural selection where the competition selects for *low* complexity of the individuals in a line of descent, not high

Complexity and the Arrow of Time, ed. Charles H. Lineweaver, Paul C. W. Davies and Michael Ruse. Published by Cambridge University Press.
© Cambridge University Press 2013.

complexity.[1] One can even engineer models where the competition ends up weeding out *all* the individuals, so that the natural selection causes all the lines of descent to die – in which case the complexity of the system has been driven to its minimal value. So natural selection, by itself, need not cause complexity to increase.

Conversely, there are biological systems that appear to increase their complexity with time but that do not involve the adaptionist process discussed above. Examples are constructive neutral evolution (Gray *et al.*, 2010), and arguably auto-catalytic systems (Kauffman, 1995), in which natural selection plays a different (and less central) role in the increase of complexity than it does in adaptionist processes.

Another example is embryogenesis of a single embryo developing in a womb; the increase in complexity of the embryo is not due to its "competing" with other embryos in any sense. At best, one might argue that the embryo's increase in complexity arose via competition occurring in the past, between its ancestors and their antagonists. This is a rather tortuous connection between a current rise in complexity of a system and the process of natural selection. It also doesn't address the possibility of hand-crafting an artificial embryo so that its complexity will rise in an artificial womb, without that womb having any ancestors.

A particularly simple kind of process that increases complexity without any natural selection is depletion forces (Asakura & Oosawa, 1958; Götzelmann *et al.*, 1998; Adams *et al.*, 1998; Marenduzzo *et al.*, 2006). As an example of such forces, consider a hollow sphere, filled with some balls that share a large size and very many balls that share a small size. Assume all interactions are simply elastic hard-sphere collisions. Start the system with the large balls uniformly distributed throughout the interior of the sphere, and the small balls uniformly distributed throughout that part of the sphere's interior

---

[1] For example, the model might have typical individuals of low complexity be more robust against external shocks than typical individuals with high complexity, which confers a selective advantage to those low complexity individuals.

where there are no large balls. Then the second law will generate "depletion forces" that drive the large balls to hug the interior wall of the enclosing sphere.[2] So the large balls will be driven from uniformly filling the interior of a sphere to uniformly filling the shell of the sphere's boundary, i.e., they will be driven to assume a more complex configuration than the one they started with. With shapes slightly more complicated than balls, far more elaborate structures can be driven to arise, e.g., sheets, helices, etc. Recent work has suggested that some of the apparent complexity of biological systems (e.g., in the internal structure of cells) arises from such processes (Marenduzzo et al., 2006).

For perhaps the most striking example of a biological system that increases in complexity without natural selection, note that, at least over certain periods and timescales, it seems that the entire terrestrial biosphere has increased its complexity, e.g., immediately following mass extinction events, or in the major life transitions. However, the "individual" of the biosphere does not undergo reproduction, and certainly is not engaged in competition with other biospheres. So external natural selection pressures cannot be the underlying cause of the increase in its complexity.[3]

Evidently then, complexity in biological systems can be driven to increase by many different processes in addition to natural selection. Since both natural selection and many of these other processes are emergent phenomena, it is natural to wonder whether there is a more fundamental and broadly applicable process that underlies and unites many of them. In particular, given the example of

---

[2] When the large balls hug the interior wall, they exclude less of the volume of the sphere's interior to the small balls, i.e., increase the available volume to the small balls. Since there are so many more small balls, this increase in available volume to the small balls more than compensates for the decrease in available volume to the large balls, and therefore increases the total Boltzmann entropy.

[3] Of course, there is natural selection occurring *within* the biosphere. But that is quite different from the way that natural selection arises in the biological examples of increasing complexity given above, where a lineage grows in complexity due to *external* natural selection pressures, not internal ones.

depletion forces, one might wonder whether the second law of thermodynamics is an underlying driver of complexity rise in many biological systems.[4] Might it somehow be that the second law, which increases disorder in a full (closed) system, not only *allows* (open) subsystems to increase their order, but actually *drives* them to do so, under certain circumstances?

11.1.1 Dynamical systems theory perspective

The goal of this chapter is to investigate this question. Ultimately, one would like to answer it in terms of dynamic systems theory. To be more precise, let $\rho(\gamma)$ refer to the phase space density of a system over phase space position γ, and let H refer to the Hamiltonian governing its dynamics. Then our goal is to understand what characteristics of H determine whether the resultant dynamics of ρ has an attractor throughout which ρ has high complexity, and to understand how the dynamics of the complexity of ρ is determined by H.

To illustrate this dynamic systems perspective in a biological context, say our system is initially described by a ρ within a basin of attraction of a high-complexity attractor. Say that the system experiences an external shock knocking it to another point in the basin that has lower complexity. After the shock, the system would start increasing its complexity back to the value it had before the shock. Examples of this arguably include asteroid impacts, volcanic eruptions, etc., that cause a mass extinction, thereby reducing the complexity of the terrestrial biosphere, after which the biosphere's complexity grows back. Note, though, that if the shock were big enough to knock the systems completely out of the basin of attraction, then the system would "die", and not increase its complexity back to what it was.

Our goal then is to investigate how the Hamiltonian of a system determines its high-complexity attractors, and in particular how its

---

[4] Of course thermodynamics is deeply involved in biochemical systems (Smith, 2008a, 2008b, 2008c). The concern here is with thermodynamic processes that are more broad-ranging.

thermodynamic properties do so. Some of the particular questions pursuant to this goal are as follows.

(1) How many separate high-complexity attractors are there for a given H?

(2) What fraction of the space of all $\rho(\gamma)$ lies in the basins of attraction of those high-complexity attractors?

(3) Are those attractors fixed points, limit cycles, strange attractors, etc.?

(4) How narrow are those basins of attraction? In other words, within a single such basin, what is the shape of the function taking complexity level χ to the volume of all $\rho(\gamma)$ within a given radius of the basin's attractor that have complexity at most equal to χ?

(5) How does the rate of increase of complexity of a particular $\rho(\gamma)$ depend on how close its complexity is to the complexity value of the attractor it is approaching?

(6) Is complexity a (negative of a) Lyapunov function of the dynamics within such a basin of attraction? In other words, is the attractor the local peak of complexity in the space of $\rho(\gamma)$? Or are there $\rho(\gamma)$ in the basin of attraction that have higher complexity than the attractor does?

(7) As an associated question, what is the maximal value of complexity per unit volume in space-time that a system can have?

(8) As an empirical issue, are almost all examples of systems that reliably increase their complexity in time biological?

(9) How big is the average high-complexity attractor of a given Hamiltonian? In other words, how much of a disruption must there be to $\rho(\gamma)$ to knock it out of the basin of attraction? (In the context of biological systems, this question amounts to asking how much of an external shock it takes to "kill" such systems.)

(10) What fraction of densities $\rho(\gamma)$ with high complexity lies in basins of a high-complexity attractor? In other words, how likely is it that a randomly chosen density with high complexity only has that high complexity temporarily?

Presumably these questions can be investigated without recourse to thermodynamics. Indeed, one might expect there to be situations where the dynamics resulting in high-complexity attractors has nothing to do with the second law. However, in this chapter I focus on situations where much about the dynamics can be understood in terms of the second law.

11.1.2 Paper roadmap

The very first step in analyzing "high-complexity attractors" is to fix a definition of "complexity". To that end, I begin in Section 11.2 by arguing in general terms that a crucial feature of complex systems is that they contain patterns that vary greatly across locations and/or scales within themselves (Wolpert & Macready, 2000, 2004, 2007). Here I quantify that variation with the Jensen–Shannon (JS) divergence among the patterns across the locations and/or scales of the system (Grosse *et al.*, 2002; Cover & Thomas, 1991; Mackay, 2003).

In the next section, I consider a particularly simple model of steady state, off-equilibrium open systems, as an archetype for systems that maintain high complexity. This model can be viewed as an abstraction of a photosynthetic organism, open to an environment of sunlight, reducing the interaction between the organism and the sunlight into what is essentially a generalization of depletion forces (Verma *et al.*, 1998; Crocker *et al.*, 1999; Marenduzzo *et al.*, 2006). I show that the second law will cause systems described by this model to stochastically grow networks, assuming certain conditions are met. (Loosely speaking, these networks can be viewed as abstractions of food webs.) I also show that such networks can be expected to have high Jensen–Shannon divergence. In this way, I show how the second law not only *allows* an open system to have high complexity, but can actually *drive* it to have high complexity.

11.2 COMPLEXITY AND JENSEN SHANNON DIVERGENCE

11.2.1 Self-dissimilarity

In almost all large systems commonly characterized as complex, the spatio-temporal patterns exhibited on different scales and/or in different regions differ markedly from one another. Conversely, for systems commonly characterized as simple the patterns are quite similar.

The human body is a familiar illustration of this; as one changes the scale of the spatio-temporal microscope with which one observes the body, the distribution of patterns that one sees varies tremendously. Similarly, as one changes from one region in the body to another, the distribution of patterns varies tremendously. The (out of equilibrium) terrestial climate system is another good illustration, having very different dynamic processes operating at all spatiotemporal scales and in different regions, and typically being viewed as quite complex. Complex human artifacts also share this property, as anyone familiar with large-scale engineering projects will attest.

The following examples describe such variations in patterns across scales and/or regions in more detail.

Example 11.1 *Consider a spatially extended physical system involving many particles. The phase space position of such a system, γ, is a mass distribution $w(r)$ across $r \in \mathbb{R}^3$, together with an associated momentum field, π. As an example, if the system is a human body, $w(r)$ would be the mass distribution of all elementary particles in that human body at a particular time.[5] In light of this meaning of γ, a phase space density $\rho(\gamma)$ evaluated at a particular time t fixes a probability density function $p(w, \pi)$ over mass distributions and associated momentum fields.*

Use i to specify the pair of a scale and location of a three-dimensional sphere, i.e., $i = (i_{scale}, i_{location})$. Define x_i for a particular i to be the possible contents of a mass distribution over r when that distribution is given by masking $w(r)$ with the sphere specified by i. So x_i is a mass density function over a Euclidean variable. To be precise, given a particular mass distribution $w(r)$, and a scale/location i, the associated x_i is defined as

$$x_i(s) = \begin{cases} w(i_{scale}s + i_{location}) & if |s| \leq d, \\ 0 & otherwise, \end{cases}$$

for some fixed window width d. Note that for any set of i's, a given $w(r)$ fixes an associated set of values $x_i(s)$, which we can represent as a vector of functions, $\bar{x}$. Therefore a distribution $p(\bar{x})$ is induced

---

[5] As discussed in Wolpert & Macready (2007), we may be interested not in the distribution $w(r)$ of all particles in the system, but only in the distribution of carbon atoms, or of water atoms, or of electrons, or some other subset of the particles of the system. For current purposes, though, we don't need to specify what types of particles $w(r)$ describes.

by any distribution p(w) (which in turn is induced by a distribution ρ(γ)).

As an illustration, we could have w(r) be the mass distribution of the human body. We could also take $d = 1$ mm, and have a set of i's sharing the value $i_{scale} = 1$, where $i_{location}$ varies across the entire human body. So the contents $x_i(s)$ would be the mass distribution w(r) masked in various ways, as described above. We could approximate $p(\vec{x})$ as a product of the distributions at the different locations, assuming those locations are sufficiently far apart, on the scale of 1 mm, so that the mass distributions (inside 1-mm-wide spheres) centered at those locations are approximately statistically independent. The mass distributions at those locations differ quite a lot from one another (assuming $i_{location}$ ranges across the many organs in the human body), so self-dissimilarity is large.

Example 11.2 *As a variation on Example 11.1, for the same human body w(r), we could have $i_{location}$ be fixed, but have i_{scale} vary across many orders of magnitude. In this case, we are interested in how much the mass distributions at a single location, but a wide range of magnifications, differ from one another. Again, for a human body, in which the distributions at the scales of organs differs drastically from that at the scale of cells, which in turn differs drastically from that at the scale of organelles, this variation is quite high.*

Example 11.3 *Another example is to use w(r) of the entire terrestrial biosphere. As with a human body, at any fixed moment in time, the mass distributions in the biosphere differ drastically from one location (i.e., within one organism) to another. Also, like with the human body w(r), mass distributions in the biosphere w(r) differ drastically from one scale to another (i.e., as one moves among levels of magnification with which a single organism is examined). So, again, the variation in patterns is quite high.*

In contrast to such complex systems, in many "simple" systems the distributions of the patterns at different scales and/or in different regions do not vary significantly from one another. For example, at scales sufficiently large compared to the size of individual molecules, the patterns at different scales and/or in different regions in a fully equilibrated ideal gas do not vary at all. Similarly, at sufficiently large scales, the patterns do not vary in a crystal.

Based on such examples, one could even argue that it is the *self-similar* aspects of simple systems, as revealed by allometric scaling, scaling analysis of networks, etc. (Stanley *et al.*, 1996), that reflects their inherently simple nature. After all, such self-similarity means that the pattern across all scales can be encoded in a short description (e.g., have low algorithmic information complexity). Such small code lengths are often taken to mean *ipso facto* that the system is not complex.

More generally, even if one could find a system commonly viewed as complex that was clearly self-similar in all important regards, it is hard to see how the same system wouldn't be considered even more "complex" if it were self-dissimilar. Indeed, it is hard to imagine a system that is highly self-dissimilar in both space and time that would not be considered complex.

Evidently, then, the *self-dissimilarity* among the patterns within many systems is an important component of their complexity (Wolpert & Macready, 2000, 2004, 2007; Parrott, 2010; Proulx & Parrott, 2008). Intuitively, the self-dissimilarity of a system reflects how much the information stored at one region or scale, and/or its processing there, differs from that at the other regions or scales. Indeed, viewing a physical system as a computational device, different regions in the system are simply different addresses, using an "addressing system" based on location. Different scales are also different addresses in the system, just using a different addressing system.

For an engineering perspective on this, note that if a system has very similar information stored at different addresses, then as an information storage device, the system is quite inefficient. It has lots of redundancy among the contents of its different addresses. Conversely, if the system has quite *different* information stored at its different addresses, then the system is efficient at encoding as much information into its state as possible. It has distributed its different parts of the information it stores into different addresses, rather than just duplicating all its information among all those addresses. This (in)efficiency is captured in the notion of self-dissimilarity.

The importance of how much the patterns in a system vary across addresses extends beyond addressing schemes based on either spatial location or spatial scale. As an illustration, complex systems often exhibit great variation in their patterns at different times, whereas simple systems typically exhibit less. For example, once it has fossilized a dead organism is static across time, i.e., completely self-similar along the time axis. What relatively little spatiotemporal complexity it still possesses is purely spatial, a relic of its complex past.

It is not being claimed that the definitive answer to the old question of what it means to say a system is "complex" is that the information at different addresses in the system varies greatly. (Ultimately, what scientists mean by the word "complex" is an issue perhaps best addressed by linguists and anthropologists.) Rather, the claim is that an important component of complexity is how distributed the information is. Furthermore, as elaborated below, this aspect of complexity turns out to be particularly well-suited to an analysis of the relation between rise in complexity and the second law.

11.2.2 Formalizing self-dissimilarity – notation

How should we formalize a system's self-dissimilarity, the amount that the "information at different addresses in the system varies"? To answer this, first we must fix some notation. Consider a vector-valued random variable, $\mathbf{X}^n$, taking values $\vec{x} \in \mathbf{X}^n$. Write a component of $\mathbf{X}^n$ as $\mathbf{X}_i$, with elements x_i. The indices i are the "addresses" referred to above, and the x_i are the contents of those addresses. Write the probability of any $\vec{x}$ as $p(\vec{x})$. Write the associated marginalizations as $P_i(x_i)$. When I want to refer to a generic $p_i(x_i)$, I will often write $p(x)$. (So the argument of p determines whether it refers to the full distribution defining $\mathbf{X}^n$ or to a generic marginalization of that full distribution.) For much of what's below, I will take each $\mathbf{X}_i$ to be the same finite space X, with $m = |X|$ elements.

Finally, write the *Shannon entropy* of $p(\vec{x})$ as

$$S_\mu(p) \equiv - \sum_{\vec{x} \in \mathrm{supp}[\mu]} p(\vec{x}) \ln\left[\frac{p(\vec{x})}{\mu(\vec{x})}\right], \qquad (11.1)$$

where, conventionally, $\mu(\vec{x})$ is viewed as a "prior probability" of $\vec{x}$. Since μ is a proper probability distribution, and so normalized, S_μ *is* non-positive, being maximized as 0 iff $p(x) = \mu(x)$.[6] Up to an irrelevant overall additive constant, the "information" in a distribution p is taken to be $-S_\mu(p)$. It is non-negative (due to the additive constant), being minimized when p is exactly what was expected *a priori*, namely μ.

An important class of scenarios is where $p(\vec{x})$ is well-approximated as a product distribution, $p(\vec{x}) = \Pi_i p_i(x_i)$. This means that the random variables at the different addresses are physically separated enough that we can treat them as decoupled. For example, in the case of a human body, with addresses being locations and the random variables being spatial configurations of amino acids, this approximation means that we consider the distribution of such configurations located in the hippocampus as statistically independent of the distribution of such configurations located in the pancreas. Note that when p is a product distribution we can write $S_\mu(p) = \sum_i \sum_{x_i \in \mathrm{supp}[\mu_i]} p_i(x_i) \ln[p_i(x_i)/\mu_i(x_i)]$.

In this chapter I restrict attention to the case where p is a product distribution. In addition, for simplicity (and to agree with much of the literature), I will take μ to be uniform. For convenience I will define

$$S(p) \equiv - \sum_{\vec{x}} p(\vec{x}) \ln[p(\vec{x})], \qquad (11.2)$$

which equals $S_\mu(p) + n \ln[m]$ evaluated under the uniform μ.

---

[6] To recover the usual formulation where entropy is *non-negative*, with a *minimal* value of 0, one adds the p-independent constant $-\min_{\vec{x} \in \mathrm{supp}[\mu]}(\ln(\mu(\vec{x})))$, to the definition of S_μ.

11.2.3 Formalizing self-dissimilarity as JS divergence

In information theory, the conventional way to quantify how much a set of distributions $p_i(x)$ vary is as their *Jensen–Shannon divergence* (JS divergence, Grosse *et al.*, 2002). Taking those p_i to be marginals of a product distribution p, JS divergence is defined as

$$JS(p) \equiv S\left(\frac{\sum_i p_i}{n}\right) - \frac{\sum_i S(p_i)}{n}. \qquad (11.3)$$

$JS(p)$ can be viewed as the average "information-theoretic distance" from a randomly chosen distribution p_i to the "center of masses" of the set of distributions $\{p_i\}$. In this sense the JS divergence can be viewed as a symmetrization and extension of Kullbach–Leibler distance (Cover & Thomas, 1991). In particular, the JS divergence equals 0 iff all of the distributions p_i are identical.

As an illustration, consider the case of a window-based addressing scheme (as in Example 11.1), where each address specifies a location, and say that we approximate p as a product distribution. Then on scales substantially larger than molecules there is *no* variability among the p_i for a perfectly equilibrated ideal gas. The same is true (at least to first order) for a perfect crystal. For such systems JS divergence equals 0. This is stated a bit more formally in the following example.

Example 11.4 *Return again to the setting for Example 11.1. For simplicity, assume the densities within each window are statistically independent, so that p is a product distribution.*

Like any other density function, p(w) contains information. In particular, we are interested in how much information in p(w) is distributed across different addresses. Physically, a large amount of information distributed across the system would mean that the patterns stored in the different addresses i typically vary a lot from one another, assuming those distributions are formed by sampling from p(w).

To see how JS divergence captures this, note that for any fixed address i, p(w) induces a distribution over x, given by

$$p_i(x) = \int dw\, p(w)\delta(x_i(s) - w(i_{\text{scale}}s + i_{\text{location}})).$$

(where both the x_i and w arguments of the Dirac delta function are being viewed as vectors with components indexed by s). Given this definition of each marginal P_i, JS(p) is very small (or zero) for some systems typically viewed as simple, like fractals, gases, and (at large enough scales) crystals. On the other hand, JS(p) is quite high for systems typically viewed as complex. For example, in a human, the distribution at any single region tends to have lots of structure, i.e., high information. Furthermore, the structure in different addresses tends to be quite distinct. The result is a high value of the difference in entropies that defines JS(p).

For another perspective on JS divergence, view the index i as a random variable with uniform prior, and identify $p(x|i) \equiv p_i(x)$. So the marginalization of $p(x, i)$ over i is just $p(x) \equiv \sum_i p_i(x)/n$, the average $p_i(x)$, and the marginalization of $p(x, i)$ over x is a uniform distribution. Then the JS divergence equals the mutual information between x and i:

$$JS(p) = S(p(x)) - \sum_i p(i)S(p(x|i))$$
$$= S(p(x)) + S(p(i)) - S(p(x, i)). \tag{11.4}$$

In other words, for a product distribution p, JS(p) quantifies how much knowing the value x of a sample of one of the distributions p_i tells you about what i is, and vice-versa. As an illustration, in Example 11.1, JS divergence is high if being provided an image in a window, where all you know about it is that it was sampled from one of the distributions of images, tells you a lot about what distribution of images it was likely sampled from. In contrast, self-dissimilarity is low if being provided an image that you know was sampled from one of the distributions of images tells you little about what distribution of images it was likely sampled from.

In Wolpert (2012) I consider the more general scenario where p need not be a product distribution. I then solve for the optimal product distribution μ. I do this using four separate arguments: based on Bayesian statistics, the maximal entropy principle, minimizing coding length, and minimizing algorithmic information complexity, all of which result in the same (non-uniform) answer for the optimal μ. Using this μ rather than the uniform μ changes the quantification of self-dissimilarity. This new quantification is the sum of JS divergence of the marginals of p and the multi-information among the associated random variables X_i. It reduces to $JS(p)$ when p is a product distribution, however, and so does not affect the analysis of this chapter, which assumes a product distribution p.

11.3 ANALYZING SYSTEMS THAT RISE IN JS DIVERGENCE

Given the choice of JS divergence to measure complexity, the next step is to investigate how the dynamics of a system's JS divergence might be related to the second law. This section presents a preliminary investigation of this issue by analyzing a simple model involving a generalized version of depletion forces.

Throughout this analysis coarse-graining will be assumed, so Liouville's theorem does not force phase space volume to be conserved. In addition, no external heat baths or any source of energy uncertainty will be assumed. So (Gibbs) physical entropy reduces to the number of distinct states available to the system.

Since the systems considered below will often be far from equilibrium, with widely varying temperatures within them, the analysis will be cast in terms of entropy gaps (between the actual entropy of a system and its maximal entropy) rather than in terms of free energy. In keeping with convention, though, rather than refering to such an "entropy gap", I will use the term "negative entropy" (negentropy) (Schrödinger, 1944; Brillouin, 1962, 1953; Mahulikar & Herwig, 2009).

11.3.1 Examples of systems that rise in JS divergence

To begin, note that many systems that appear to have an attractor with high complexity are thermodynamically open. In such cases there is actually a composite system $A \times B$ undergoing Hamiltonian dynamics. It is the dynamics of B that, evolving under the external influence of A, has an attractor with high complexity. In such systems, loosely speaking, A is a negative-entropy (negentropy) flux flowing from some low-entropy third system T. B "harvests" that flux, and by doing so B increases its complexity. Some examples of this are:

(I) the terrestrial biosphere is B, robustly maintaining a high complexity by harvesting the negentropy flux of sunlight (which plays the role of A) flowing from the Sun (which plays the role of T);

(II) an ecosystem at a hydrothermal vent is B, robustly maintaining a high complexity by harvesting the negentropy flux of the flow through the vent (which plays the role of A) from within the Earth (which plays the role of T);

(III) lineages in a bird species whose members regularly ride atmospheric drafts, and in a plant species whose members regularly release fertilized seeds on the wind, which are examples of systems B that harvest the negentropy flux of the wind (which plays the role of A) to maintain a high complexity (i.e., to live and reproduce), with sunlight and the Earth's rotation playing the role of T (since they are what drives the wind);

(IV) organisms resident in a fixed region in a stream that use that stream's current to flush waste are examples of systems B that harvest the negentropy flux of the stream A to maintain a high complexity (i.e., to live).

Note that in all these examples the system B has relatively high JS divergence in its attractor state, e.g., using the addressing scheme in Ex.1.

In addition to the broader questions raised in Section 11.1, these examples raise many questions of their own. For example, in all these examples, much of the negentropy flux is not harvested. Is that always the case? Is there an upper bound to what fraction of the flux can be harvested for an extended period of time by a system with an attractor of high complexity? Is such harvesting of negentropy flux the only way that a system can have an attractor with high complexity? Or can thermodynamically closed systems have such attractors as well? A final example of such questions is whether the existence of such attractors of high JS divergence in these examples is a result of the second law somehow. This question forms the basis for the rest of the analysis of this section.[7]

11.3.2 JS divergence and off-equilibrium steady state systems
In all of the examples of Section 3.1, the attractor state is

(1) stationary (steady state);
(2) stable against small perturbations;
(3) out of equilibrium (and therefore, given (1), in contact with an external environment);
(4) able to quickly move to an equilibrium if the system is isolated from its environment.

For example, the terrestrial biosphere is (over certain timescales) in a steady state, with many of its large-scale physical characteristics stable against small perturbations. It is also out of equilibrium, and if it were isolated from its environment (i.e., were deprived of sunlight and hydrothermal vent effluent) would quickly move to equilibrium (i.e., all life would cease).

I call states with properties (1)–(4) *vital* states.[8] Intuitively, systems in a vital state are exploiting their environment to stably maintain a far from equilibrium configuration, despite strong thermodynamic "forces" pushing them to equilibrium. The analysis of such systems and their (non-equilibrium) thermodynamics is a deep and mature field with roots going back decades (Onsager 1931a, 1931b; Prigogine, 1945; Grandy, 2008; Lebon et al., 2008). Here I will only be

---

[7] This is similar to the question of whether an external flux of negentropy drives the emergence of life. A seminal analysis of this issue can be found in Morowitz & Smith (2007).

[8] Note that each of properties (1), (2), and (4) have an implicit timescale; I will take this timescale to be the same for all of them.

interested in sketching some relevant properties of such systems, to allow them to be related to the second law and JS divergence.

Since a vital state is stable against small perturbations, it is an attractor. By hypothesis it is also off-equilibrium, when considered separately from its environment. I assume that to be such an off-equilibrium attractor, a vital state cannot dissipate entropy to its environment (and thereby approach equilibrium) via internal thermalization faster than a rate σ. (In particular, it cannot dissipate entropy faster than it acquires negentropy from the environment.) In addition to holding in many biological systems, this *thermalization rate limit* seems to hold in many artificial complex systems. For example, it is very difficult to make large computers dissipate heat beyond a certain rate. (This is why it is extremely important to design large computers so that they do not generate heat too quickly.)

11.3.3 Vital states formed by coupling two subsystems

I now consider vital systems that comprise one or more pairs of coupled subsystems. Each of those subsystems will correspond to a separate "address" of the system, in the sense of Section 11.2.3. So I will assume that probability distributions over possible states of the joint system are well-approximated by a product distribution over states of the constituent subsystems (at least at appropriate moments in time).

Vital systems that comprise more than one pair of coupled subsystems often have an elaborate network structure linking the pairs. There has been a lot of analysis done on these kinds of networks (e.g., see Morowitz (1968), Ho (1994), and other work analyzing free energy flows and free energy networks in ecosystems). However for current purposes I only need to consider very simple aspects of such systems.

I start with an analysis for systems comprising a single pair of subsystems. Let A be the set of states a of a "background" subsystem $\mathcal{A}$, and let B be the set of states b of a "foreground" subsystem $\mathcal{B}$, where for simplicity I restrict attention to the case where both A and B are finite. Assume that during a time interval $[\tau, \tau + \delta t]$ the subsystems $\mathcal{A}$ and $\mathcal{B}$ are coupled, and that there are no other systems interacting with

A and/or B during that interval. To ground intuition, B can be a (highly abstracted) model of a population of photosynthesizing organisms, or even of a photovoltaic cell, and A can be a (highly abstracted) model of some sunlight incident upon that population during the interval $[\tau, \tau + \delta t]$.

Let $N_{AB}(t)$ be the number of joint states $\{ab : a \in A, b \in B\}$ that can occur at time t.[9] Assume that those states are indistinguishable to an external observer. Let $N_A(t)$ be the number of states in A that can occur at t, again assuming that they are indistinguishable to an external observer. Define $N_B(t)$ similarly. Note that $N_{AB}(t) \le N_A(t)N_B(t)$.

Assume that, due to the Hamiltonian governing the interaction of the two subsystems, the following two conditions hold:

(i) $N_B(\tau) > N_B(\tau + \delta t)$,
(ii) $N_A(\tau)N_B(\tau) > N_{AB}(\tau + \delta t)$.

Assumption (i) means that the coupled system ends the interval with lower Boltzmann entropy for B. However, assumption (ii) means that the coupling increases total Boltzmann entropy. (Recall that we are assuming coarse-graining, so that available phase space volume need not be conserved.)

As a simple example, both assumptions (i) and (ii) hold with systems being subject to depletion forces, where B refers to the larger balls, A to the far more numerous tiny balls, τ is the initial time when the balls in B are uniformly distributed throughout the interior of the sphere, and $\tau + \delta t$ is the time after the system has reached equilibrium with the balls B hugging the inner wall of the sphere. Note that the fine-grained states a and b are highly correlated at $\tau + \delta t$. But, due to coarse-graining, we do not see any of that correlation. This justifies the approximation that we describe the distribution over the joint system as a product distribution.

---

[9] These can be the number of distinct eigenstates if A and B are quantum mechanical systems, or for classical systems, they can be the number of coarse-grained bins of phase space.

Note that assumption (i) would be violated if A were a heat bath for B and we used fine-grained states (rather than coarse-grained). This is because in such a case the "coupling" between the two subsystems is simply energy transfer between them. More complicated coupling is needed for assumption (i) to hold, coupling that relates the configuration of A to that of B. (In the case of depletion forces coupling, where (i) holds, no particle ever changes its energy, in contrast to the case of heat bath coupling.)

It is always true that

$$\min\{N_A(\tau + \delta t), N_B(\tau + \delta t)\} \le N_{AB}(\tau + \delta t)$$
$$\le N_A(\tau + \delta t)N_B(\tau + \delta t). \qquad (11.5)$$

(As a simple example, in depletion forces the second bound in Eq. (11.5) holds exactly: $N_{AB}(\tau + \delta t) = N_A(\tau + \delta t)N_B(\tau + \delta t)$.) When assumptions (i) and (ii) hold though, by plugging (i) into (ii) we see that

$$N_B(\tau + \delta t)N_A(\tau) < N_{AB}(\tau + \delta t). \qquad (11.6)$$

If we now plug the second inequality in Eq. (11.5) into Eq. (11.6), we see that when (i) and (ii) hold, $N_A(\tau) < N_A(\tau + \delta t)$. Indeed, plugging assumption (ii) into the second inequality in Eq. (11.5) gives

$$\ln\left[\frac{N_A(\tau)}{N_A(\tau + \delta t)}\right] < \ln\left[\frac{N_B(\tau + \delta t)}{N_B(\tau)}\right]. \qquad (11.7)$$

So the shrinkage in the entropy of B is more than offset by the growth in the entropy of A.

Whenever assumptions (i) and (ii) hold I say that A is *harvested* by B. I define the *harvest rate* of the coupling during $[T, \tau + \delta t]$ as

$$\varepsilon(\tau) = \frac{\ln\left[\dfrac{N_{AB_{(\tau+\delta t)}}}{N_A(\tau)N_B(\tau)}\right]}{\delta t}$$
$$= \frac{\ln[N_{AB}(\tau + \delta t)] - \ln[N_A(\tau)] - \ln[N_B(\tau)]}{\delta t}, \qquad (11.8)$$

where δt is implicit. The *instantaneous harvest rate* is defined as $\lim_{\delta t \to 0} \varepsilon(\tau)$. By assumption (ii), $\varepsilon(\tau)$ is positive, reflecting the fact that entropy increases during the coupling.

We can construct a long-lasting, stationary off-equilibrium version of this harvesting process by chaining instances of it one after the other. In the first step of such a process, the harvest described above occurs, in a time interval $[\tau, \tau + \delta t_1]$. Next, during $[\tau + \delta t_1, \tau + \delta t_1 + \delta t_2]$, two things happen. First, A and B decouple from each other, without increasing N_A or N_B. For simplicity I restrict attention to cases where, once the decoupling has occurred, the number of possible states of A is $N_A(\tau + \delta t_1)$ and the number of possible states of B is $N_B(\tau + \delta t_1)$. This means that the subsystems are made statistically independent in the decoupling. This is in keeping with our presumption that the probability distribution over the joint system is a product distribution at appropriate moments in time.

Next one of two essentially equivalent processes occurs.

(1) B and A get reset to the states they were in at τ. One way for this to occur is for the entropy of B to increase due to the second law while the entropy of A decreases due to its coupling to an external system.

(2) B gets reset this way. However, A leaves the picture, and is replaced by a new subsystem that is identical to A as it was at time τ.

As an example, the second process is what happens when B is a photovoltaic cell "harvesting negentropy flux" in the form of sets of photons A streaming by.

Whichever two-step process we use to return the joint system to its initial state, the total amount of entropy gained is the amount of negative entropy that had to be introduced at the end of the process to return the joint system to its starting state. This equals

$$(\ln[N_A(\tau + \delta t_1)] - \ln[N_A(\tau)]) + (\ln[N_B(\tau + \delta t_1)] - \ln[N_B(\tau)]).$$

$$(11.9)$$

Note that the first term inside the parentheses is positive while the second is negative. However, by Eq. (11.7) the magnitude of the first term is larger. So this total gain of entropy – the total negentropy harvested – is positive.

Note also that the entropy dissipated by thermalization in the second step of the two-step process is $\ln[N_B(\tau)] - \ln[N_B(\tau + \delta t_1)]$. So by the thermalization rate limit,

$$\ln[N_B(\tau)] - \ln[N_B(\tau + \delta t_1)] \leq \sigma \delta t_2. \tag{11.10}$$

This gives a lower bound on how quick the second step can be, and therefore an upper bound on the harvest rate:

$$\varepsilon(\tau) = \frac{(\ln[N_A(\tau + \delta t_1)] - \ln[N_A(\tau)]) + (\ln[N_B(\tau + \delta t_1)] - \ln[N_B(\tau)])}{\delta t_1 + \delta t_2}$$

$$\leq \frac{(\ln[N_A(\tau + \delta t_1)] - \ln[N_A(\tau)]) + (\ln[N_B(\tau + \delta t_1)] - \ln[N_B(\tau)])}{\delta t_1 - (\ln[N_B(\tau + \delta t_1)] - \ln[N_B(\tau)])/\sigma}$$

$$\tag{11.11}$$

Consider the case where A can either go through this two-step process, or can never couple to B, simply thermalizing by itself. Of the two options, the two-step process will be thermodynamically preferred by the second law if the harvest rate is greater than the thermalization limit of A, since under those conditions entropy will be higher at the end of $\tau + \delta t_1 + \delta t_2$ if the coupling occurs than if it does not. This can be illustrated by the example mentioned above where A is sunlight and B is a vastly simplified version of photosynthetic organisms. Say the photosynthetic organisms, with their thermalization rates, can harvest the negentropy of A faster (as determined by Eq. (11.11) than A can lose that negentropy by itself. Then A will couple to B. Otherwise the light will not couple to B this way, and will simply pass through or reflect off of B.

11.3.4 Vital states formed by coupling more than two subsystems

We can construct a long-lasting, stationary off-equilibrium version of this process that involves more than two steps. As an example, say that at time $\tau + \delta t_1$, B gets decoupled from A, as before. However, rather than have the two-step process repeat, B gets coupled to a third subsystem C. Then during $[\tau + \delta t_1, \tau + \delta t_1 + \delta t_2]$, rather than get reset, B itself gets harvested, by C, in a process that obeys assumptions (i) and (ii). Presume that this causes the entropy of B to increase back to the value it had at τ. Furthermore, assume that, just as in the two-step process, in the three-step process, by the time $\tau + \delta t_1 + \delta t_2$, the subsystem A gets reset to its time-τ state (or is replaced by a subsystem whose state at $\tau + \delta t_1 + \delta t_2$ is identical to what the state of A was at time τ). Finally, assume that in an interval $[\tau + \delta t_1 + \delta t_2, \tau + \delta t_1 + \delta t_2 + \delta t_3]$, C has its entropy increase to the value it had at time $\tau + \delta t_1$. So by time $\tau + \delta t_1 + \delta t_2 + \delta t_3$ the entire three-step process has returned the joint system to the original joint state it had at τ.

In this three-step process the total entropy gained is

$$
\ln\left(\left[N_A(\tau + \delta t_1)\right] - \ln[N_A(\tau)]\right)
$$
$$
+ \left(\ln[N_C(\tau + \delta t_1 + \delta t_2)] - \ln[N_C(\tau + \delta t_1)]\right). \tag{11.12}
$$

The second term in the parentheses in Eq. (11.12) has smaller magnitude than the second term in the parentheses in Eq. (11.9). Accordingly, more entropy is gained in this three-step process than in the two-step process it starts with. However the three-step process also takes more time than the two-step process. So to see if it is thermodynamically preferred to attach C to A B – to compare the highest possible harvest rates of the two-step and three-step processes – we have to consider the time lengths of those two processes.

To do this, first note that in the three-step process we never need to thermalize B. So the thermalization rate limit B does not provide bounds on δt_2. Instead we need to thermalize C so that the thermalization rate limit of C applies to δt_3:

$$
\ln[N_C(\tau + \delta t_1)] - \ln[N_C(\tau + \delta t_1 + \delta t_2)] \leq \sigma_C \delta t_3. \tag{11.13}
$$

Combining, we see that the harvest rate for the three-step process is

$$
\varepsilon(\tau) = \frac{\left(\ln[N_A(\tau + \delta t_1)] - \ln[N_A(\tau)]\right) + \left(\ln[N_C(\tau + \delta t_1 + \delta t_2)] - \ln[N_C(\tau + \delta t_1)]\right)}{\delta t_1 + \delta t_2 + \delta t_3}
$$
$$
\leq \frac{\left(\ln[N_A(\tau + \delta t_1)] - \ln[N_A(\tau)]\right) + \left(\ln[N_C(\tau + \delta t_1 + \delta t_2)] - \ln[N_C(\tau + \delta t_1)]\right)}{\delta t_1 + \delta t_2 - \left(\ln[N_C(\tau + \delta t_1 + \delta t_2)] - \ln[N_C(\tau + \delta t_1)]\right)/\sigma_C}
$$
$$
\tag{11.14}
$$

The δt_2 in Eq. (11.14) does not need to be long enough for B to be thermalized in the three-step process, and therefore is not limited by the thermalization rate limit of B. So it can be shorter than the δt_2 occurring in the two-step process. The term $\ln[N_C(\tau + \delta t_1 + \delta t_2)] - \ln[N_C(\tau + \delta t_1)]$ in Eq. (11.14) is negative, just like the analogous term $\ln[N_B(\tau + \delta t_1)] - \ln[N_B(\tau)]$ from Eq. (11.11). However, it has smaller magnitude. Therefore the numerator in Eq. (11.14) is larger than the numerator in Eq. (11.11). In addition, up to the term δt_2, the denominator in Eq. (11.14) is smaller than the numerator in Eq. (11.11). Therefore, whether the two-step or three-step process has a higher harvest rate – and is therefore thermodynamically preferred – is determined by the size of the δt_2s in the two-step and three-step processes.

This means that for an appropriately small three-step δt_2 compared to the two-step δt_2, if a composite subsystem $A\,B$ joined in a two-step process encounters a subsystem C, it is thermodynamically preferable for $A\,B$ to couple to C at its "tail" B, rather than stay isolated from C. So the second law will induce such coupling into a three-step process whenever the opportunity arises.

This phenomenon provides a general way for the second law to create chains of many steps linking subsystems that have vital states. To illustrate this, recall the example discussed above where A is sunlight and B is a population of photosynthesizing organisms. In this case an example of a system C is a (highly abstracted) population of herbivores. They increase the entropy of the photosynthesizing organisms by digesting them, and that increase occurs more quickly than it does if instead those organisms increase their entropy by themselves, in a steady state without any herbivores.

11.3.5 Vital states formed when more than two subsystems are coupled simultaneously

There is no *a priori* restriction that negentropy harvests can only involve coupling two subsystems at once. There are many extensions of assumptions (i) and (ii) to concern tuples of more than two subsystems, in which a "foreground" system loses entropy even though the total system gains entropy.

Since these kinds of harvest involve more than two subsystems being coupled, in principle multiple instances of them can be joined to build joint systems with vital states that are more elaborate than chains. Most generally, sets of harvests involving such tuples of subsystems can build systems that are graphs. In particular, whereas harvests involving two subsystems at a time cannot result in cycles, harvests involving more than two subsystems at once can result in an arbitrary number of cycles.

This is what often happens in a real biological system. To give an intuitive example, there are many organs in a human that need to all be functioning for the body as a whole to maintain homeostasis (i.e., to maintain a vital state). Those organs do not interact only pairwise. Typically, any single organ needs to be interacting with multiple other organs, in an elaborate graph, for the body as a whole to maintain homeostasis.

11.4 VITAL STATES AND JS DIVERGENCE

In all negentropy harvesting systems, the second law "glues" the subsystems together. In general it can only glue such subsystems together if

(1) their interaction Hamiltonian obeys assumptions (i) and (ii) (or more generally, the interaction Hamiltonian of whatever tuples of subsystems are available obeys the generalizations of those assumptions);
(2) the thermalization rate limits and associated harvest times are related by the equations discussed in Sections 11.3.3 and 11.3.4.

Only those subsystems that are "available" in that they obey these conditions can be glued together this way by the second law. In general, a stochastic process creates the subsystems that are available in this sense and so can be glued together. (In the example of the biosphere, depending on the timescale at which one models the biosphere, the stochastic process can involve anything from mutation and cross-over to the major transitions in life.) *A priori*, one would not expect such a stochastic process to join subsystems in a way that makes them have similar configurations. Therefore one would expect it to create overall systems with vital states that have relatively high JS divergence among their subsystems.

This relationship between the second law and complexity is a subtle one. The second law, operating through a stochastic process, "glues together" subsystems, to make a full system that has a vital state. In turn, that full system is likely to have high JS divergence, simply because there is no *a priori* reason for the constituent subsystems to have similar configurations in the vital state. In this sense, it is almost accidental that systems with vital states – attractors of the underlying dynamics that arise due to the second law – are complex.

There is a long-standing debate in the literature about whether increasing complexity in the biosphere has been driven or is a simple drift process starting from an initial condition of low complexity (Carroll, 2001; Krakauer, 2011). In the process elucidated in this chapter, the second law is "driving" the process of gluing each new edge in a harvest network. However, the introduction of nodes to be glued together with an edge is instead determined by a "drift process". In this sense, both sides of the debate are correct.

11.5 FUTURE WORK

The analysis in this chapter shows how the second law can drive the construction of dynamical attractors that have high values of such complexity. While necessary, this leaves all of the questions in Sections 11.1.1 and 11.3.1 yet to be analyzed. In addition, even the analysis that *is* done in this chapter is far from complete. To give one of the most obvious examples, how should we choose the coordinate system with which to decompose x? Given that we want $\mu(x)$ to be a product distribution over that coordinate system, a natural set of ways to choose the coordinate system is to apply principle components analysis (when x is a Euclidean vector), or a more sophisticated technique like independent components analysis. It remains to be seen, though, whether such approaches result in reasonable measures of self-dissimilarity.

As another example of a question that needs to be analyzed, as mentioned above, assumptions (i) and (ii) need to be extended to allow negentropy harvests that involve more than two interacting subsystems at once. Which such extensions are best suited to modeling the kinds of harvests that seem to occur in real-world biological systems? As another example of an important question not yet analyzed, can we bound the total harvest efficiency in a sequence of harvest links in terms of the efficiencies of each individual link? If so, does this provide an experimentally testable claim? (For example, it might mean that entropy increases faster when a plant is coupled to light, water, CO_2, etc., than when it is decoupled from them (and therefore starts to die).)

Another set of issues concern the vaguely described "stochastic processes" under which subsystems arise that jointly meet assumptions (i) and (ii) (and their extensions), and therefore can be glued together by the second law. What are the characteristics of such processes (biological or otherwise) in the real world? For example, are random growth networks appropriate models? Whatever form they take, do the stochastic process models allow us to relate harvest rates, the size of the harvest network, and JS divergence? (Note that until something precise is said about the stochastic process, little quantitative can be said concerning the probability distribution over possible values of JS divergence.) More generally, does the form of such processes provide an explanation of the many power laws that govern the biosphere, food webs, etc. (e.g., power laws of complexity vs total biomass)?

There are several important questions to ask concerning any particular stochastic growth model of harvest nets. For example, given such a model, what is the associated probability of arising at a net with overall rate ε, and how does that compare to the ε distribution induced by uniformly sampling all nets? A related question is whether under some such growth model, any network G_1 with overall harvest rate $\varepsilon_1 > \varepsilon_2$ is *always* thermodynamically preferred to a network G_2 with overall rate ε_2 (and therefore has higher probability). One of the factors that goes into answering this question is whether there can be a subnet of G_1 that has a harvest rate smaller than any of the subnets of G_2, even though G_1 has a higher overall harvest rate.[10]

If, under some given growth models, nets do get more probable as their harvest rate increases, independent of the harvest rates of their subnets, then even something like a human city, which is a huge network, is driven to have a high harvest rate. This would be rather astonishing, given that one normally models how complexity forms in cities in terms of human–human cognitive interactions, not in terms of anything so physical as negentropy.

A related (empirical) question is whether the rise of human civilization, including all the complexity of human cultures, resulted in a growth or loss of self-dissimilarity of the entire biosphere.[11]

All of these are issues for future research.

11.6 Summary

Many systems appear to increase their "complexity" in time and then robustly maintain a high complexity once achieved (Krakauer, 2011; Carroll, 2001; McShea, 1991; Smith, 1970). To investigate this phenomenon it is necessary to formalize "complexity". Here I have built on recent work arguing that complexity of a system should be formalized as to how much the patterns exhibited on different scales and/or at different locations of that system differ from one another. I quantified this variation in patterns – this type of complexity – as the Jensen–Shannon (JS) divergence among the patterns.

---

[10] A literature tangentially related to such modeling concerns the "maximum entropy production principle" (Paltridge, 1979; Martyushev & Seleznev, 2006; Grinstein & Linsker, 2007). This rather controversial literature traces its lineage back to Onsager (1931a).

[11] One would likely want to analyze this question using an extension of this paper's quantification of self-dissimilarity, since the analysis would concern *p*s that are not product distributions. For example, one might want to use the extension of self-dissimilarity that is discussed in Wolpert (2012).

Next I constructed a highly stylized model of off-equilibrium, steady-state, network systems whose structure is maintained by depletion forces. Such networks can be viewed as highly abstracted models of living systems (organisms, ecosystems, or entire biospheres), bypassing considerations of reproduction and natural selection to focus on the underlying physics and information theory.

Finally, I showed how the second law can drive the growth of these depletion force network systems. I also showed that this growth causes such networks to have high JS divergence. In this way the second law can actually drive the increase of complexity in time.

Acknowledgements

I would like to thank Giovanni Bellesia for very stimulating discussions.

REFERENCES

Adams, M., Dogic, Z., Keller, S., & Fraden, S. (1998). Entropically driven microphase transitions in mixtures of colloidal rods and spheres. *Nature*, **393**, 349–352.

Asakura, S. & Oosawa, F. (1958). *J. Polym. Sci.*, **33**, 183.

Brillouin, L. (1953). Negentropy principle of information. *Journal of Applied Physics*, **24**, 1152–1163.

Brillouin, L. (1962). *Science and Information Theory*. New York: Academic Press.

Carroll, S. (2001). Chance and necessity: the evolution of morphological complexity and diversity. *Nature*, **409**, 1102–1109.

Cover, T. & Thomas, J. (1991). *Elements of Information Theory*. New York: Wiley-Interscience.

Crocker, J. C., Matteo, J. A., Dinsmore, A. D., & Yodh, A. G. (1999). Entropic attraction and repulsion in binary colloids probed with a line optical tweezer. *Physical Review Letters*, **82**, 4352–4355. doi/10.1103/PhysRevLett.82.4352.

Götzelmann, B., Evans, R., & Dietrich, S. (1998). Depletion forces in fluids. *Phys. Rev. E*, **57**, 6785–6800. doi/10.1103/PhysRevE.57.6785.

Grandy, W. T., J. (2008). *Entropy and the Time Evolution of Macroscopic Systems*. Oxford: Oxford University Press.

Gray, M. W., Luke, J., Archibald, J. M., Keeling, P. J., & Doolittle, W. F. (2010). Irremediable complexity? *Science*, **330**(6006), 920–921.

Grinstein, G. & Linsker, R. (2007). Comments on a derivation and application of the 'maximum entropy production' principle. *J. Phys. A: Math. Theor.*, **40**, 971720.

Grosse, I., Bernaola-Galvan, P., Carpena, P., Roman-Roldan, R., Oliver, J., & Stanley, H. E. (2002). Analysis of symbolic sequences using the Jensen–Shannon divergence measure. *Phys. Rev. E*, **65**, 041905.

Ho, M. (1994). What is negentropy? *Modern Trends in BioThermoKinetics*, **3**, 50–61.

Kauffman, S. A. (1995). *At Home in the Universe: the Search for the Laws of Self-Organization and Complexity*. New York: Oxford University Press.

Krakauer, D. (2011). Darwinian demons, evolutionary complexity, and information maximization. *Chaos*. **21**, 037110, dx.doi.org/10.1063/1.3643064.

Lebon, G., Jou, D., & Casas-Vazquez, J. (2008). *Understanding Non-equilibrium Thermodynamics: Foundations, Applications, Frontiers*. New York: Springer-Verlag.

Mackay, D. (2003). *Information Theory, Inference, and Learning Algorithms*. Cambridge: Cambridge University Press.

Mahulikar, S. & Herwig, H. (2009). Exact thermodynamic principles for dynamic order existence and evolution in chaos. *Chaos, Solitons and Fractals*, **41**, 1939–1948.

Marenduzzo, D., Finan, K., & Cook, P. (2006). The depletion attraction: an under-appreciated force driving cellular organization. *Journal of Cell Biology*, **5**, 681–686.

Martyushev, L. & Seleznev, V. (2006). Maximum entropy production principle in physics, chemistry and biology. *Physics Reports*, **426**, 1–45.

McShea, D. (1991). Complexity and evolution: what everybody knows. *Biology and Philosophy*, **6**, 303–324.

Morowitz, H. (1968). *Energy Flow in Biology*. Waltham, MA: Academic Press.

Morowitz, H. & Smith, E. (2007). Energy flow and the organization of life. *Complexity*, **13**, 51–59.

Onsager, L. (1931a). Reciprocal relations in irreversible processes: 1. *Physical Review*, **37**, 405.

Onsager, L. (1931b). Reciprocal relations in irreversible processes: 2. *Physical Review*, **38**, 2265.

Paltridge, G. (1979). Climate and thermodynamic systems of maximum dissipation. *Nature*, **279**, 5714.

Parrott, L. (2010). Measuring ecological complexity. *Ecological Indicators*, **10**(6), 1069–1076.

Prigogine, I. (1945). Moderation et transformations irreversibles des systemes ouverts. *Bulletin de la Classe des Sciences, Académie Royale de Belgique*, **31**, 600–606.

Proulx, R. & Parrott, L. (2008). Measures of structural complexity in digital images for monitoring the ecological signature of an old-growth forest ecosystem. *Ecological Indicators*, **8**(3), 270–284.

Schrödinger, E. (1944). *What is Life?* Cambridge: Cambridge University Press.

Smith, E. (2008a). Thermodynamics of natural selection i: Energy flow and the limits on organization. *Journal of Theoretical Biology*, **252**(2), 185–197.

Smith, E. (2008b). Thermodynamics of natural selection ii: Chemical carnot cycles. *Journal of Theoretical Biology*, **252**(2), 198–212.

Smith, E. (2008c). Thermodynamics of natural selection iii: Landauer's principle in computation and chemistry. *Journal of Theoretical Biology*, **252**(2), 213–220.

Smith, J. M. (1970). Time in the evolutionary process. *Studium Generale*, **23**, 266–272.

Stanley, M. H. R., Amaral, L. A. N., Buldyrev, S. Y. *et al.* (1996). Scaling behaviour in the growth of companies. *Nature*, **379**, 804–806.

Verma, R., Crocker, J. C., Lubensky, T. C., & Yodh, A. G. (1998). Entropie colloidal interactions in concentrated dna solutions. *Physics Review Letters*, **81**, 4004–4007, doi/10.1103/PhysRevLett.81.4004.

Wolpert, D. (2012). How to salvage Shannon entropy as a complexity measure. *Submitted*.

Wolpert, D. H. & Macready, W. (2000). Self-dissimilarity: an empirically observable measure of complexity. In Y. Bar-Yam (ed.), *Unifying Themes in Complex Systems*. New York: Perseus Books, pp. 626–643.

Wolpert, D. H. & Macready, W. (2004). Self-dissimilarity as a high dimensional complexity measure. In Y. Bar-Yam (ed.), *Proceedings of the Fifth International Conference on Complex Systems*. New York: Perseus Books.

Wolpert, D. H. & Macready, W. (2007). Using self-dissimilarity to quantify complexity. *Complexity*, **12**, 77–85.

Part IV **Philosophical perspectives**

12 Wrestling with biological complexity: from Darwin to Dawkins

Michael Ruse

> The inhabitants of each successive period in the world's history
> have beaten their predecessors in the race for life, and are, in so far,
> higher in the scale of nature; and this may account for that vague yet
> ill-defined sentiment, felt by many palæontologists, that organisation
> on the whole has progressed.

(Darwin, 1859, 345)

I have long been rather dubious about the notion of complexity. Certainly, it is not something that is always desirable. For instance, in theology, as is well known, a central Christian claim is that God is the ultimate simple. In mathematics, simplicity and elegance are highly valued. One thinks for example of the Euler identity: $e^{i\pi} + 1 = 0$. And in science too, there is much to be said for simplicity. Charles Darwin's supporter Thomas Henry Huxley, on being informed about natural selection, is supposed to have exclaimed: "How simple, how stupid not to have thought of that!" More generally in science, the ultimate in scientific achievement is bringing disparate parts of enquiry together under the same hypothesis. This process, known as a "consilience of inductions", is the paradigmatic exercise in producing simplicity (Whewell, 1840).

However, there is no doubt that the notion of complexity (and related notions of progress) continue to fascinate and absorb our attention. In this discussion, I would like to offer a few remarks about complexity (and progress) as they play out in evolutionary biology. To anticipate, it has long been recognized by historians of evolutionary

Complexity and the Arrow of Time, ed. Charles H. Lineweaver, Paul C. W. Davies and Michael Ruse. Published by Cambridge University Press.
© Cambridge University Press 2013.

biology that in the middle of the nineteenth century in Britain there were two competing views of the evolutionary process (Richards, 1987; Ruse, 2013a). One was expressed by Charles Darwin in his *Origin of Species* (1859), the other by Herbert Spencer in a series of writings, culminating in the 1860s with his *First Principles* (1862) and his *Principles of Biology* (1864). Until recently, historians were inclined to downplay the significance of Spencer. Even today he is rather considered an exemplar of all that was wrong with Victorian thinking. However, all will acknowledge that at the time Spencer was as popular as if not more popular than Darwin. And today there is growing recognition that Spencer's influence was lasting and perhaps we are even still wrestling with his legacy (Ruse, 2013a). It is my claim that the difference between Darwin and Spencer is in major respects epitomized by their differences over the notion of complexity, and that these differences can indeed be found in the thinking of biologists and philosophers of our time. I don't want to make generalizations that are too sweeping, but I would suggest that perhaps Darwin's thinking is more common in Britain and Spencer's thinking more common in the United States. This was certainly the case in the Victorian era.

12.1 CHARLES DARWIN

Let me start with a number of generally accepted facts. First, as the quote from the *Origin* just above states clearly, by and large evolutionists – including Darwinian evolutionists – think that there is some kind of development in the history of life, from the simple to the complex (Ruse, 1996). Call it "progress" if you will, but it does seem true. More so really than in Darwin's day because he knew nothing of the pre-Cambrian. But if you go back over three billion years, to the earliest life forms, at least at the physical level there is surely little question that in some way they are not as sophisticated as elephants and redwoods, eagles and dandelions – and humans. Second, it really is very difficult to pin down exactly what is meant by "progressive organization", to use Darwin's term. What does it mean to say that there has been progress to the more complex? Third, whatever

Darwin may have said about successors "beating" predecessors in the struggle for existence, in some deep sense Darwinism is corrosive towards the whole idea of upwards change. It is no tautology, in the sense that it simply defines what is success in terms of success. But it is relativistic. Success in one situation may not be success in another situation. What might be sauce for the goose is by no means always sauce for the gander (Ruse, 2006). A clever adaptation in one situation may be too expensive or too liable to break down in another situation. Sometimes the best strategy is that old saw: "Keep it simple, stupid". At times like this I like to quote the paleontologist, the late Jack Sepkoski, talking about the virtues of intelligence – something requiring complexity if anything does. "I see intelligence as just one of a variety of adaptations among tetrapods for survival. Running fast in a herd while being as dumb as shit, I think, is a very good adaptation for survival" (Ruse, 1996, 486).

So, where do we go from here? Let us see what Darwin himself had to say. Above all, we have to recognize that Darwin wrestled with the notions of complexity and progress, clearly thinking that in some respects progress involves a rise in complexity; although at the same time it wasn't just complexity for its own sake, but rather because in some respects complex organisms are "better" than simple organisms, whatever that might mean. (One thing that it obviously would mean – one thing that it obviously would mean for Darwin – would be "human-like" in some sense.) At the end of 1858, in other words in the middle of the time that he was writing the *Origin*, Darwin wrote to his good friend Joseph Hooker on the subject, showing that he was thinking about complexity and progress but was still somewhat at sea. I quote the first part, which deals directly with our topics. Even so, this is still a long quote, but I want to make my point – namely that he wrestled with the problem.

> Your letter has interested me *greatly*; but how inextricable are
> the subjects, which we are discussing. – I do not think I said
> that I thought the productions of Asia were *higher* than those of
> Australia. I intend carefully to avoid this expression, for I do not

think that any one has a definite idea what is meant by higher, except in classes which can loosely be compared with man. On our theory of Nat: Selection, if the organisms of any area belonging to the Eocene or Secondary periods, were put into competition with those now existing in the same area (or probably in any part of world) they (i.e. the old one) would be beaten hollow and be exterminated; if the theory be true, this must be so. In same manner I believe a greater number of the productions of Asia, the largest territory in the world, would beat those of Australia, than conversely. So it *seems* to be between Europe and N. America, for I can hardly believe in the difference of stream of commerce, causing so great a difference in proportions of immigrants [from East to West as opposed to from West to East].

But this sort of highness (I wish I could invent some expression, and must try to do so) is different from highness in the common acceptation of the word. It might be connected with degradation of organisation; thus the blind degraded worm-like snake (*Typhlops*) might supplant the true earth-worm; here then would be degradation in the class, but certainly increase in the scale of organisation in the general inhabitants of the country: on the other hand it would be quite as easy to believe that true earth-worms might beat out the *Typhlops*. – I do not see how this "competitive highness" can be tested in any way by us. And this is a comfort to me when mentally comparing the Silurian and Recent organisms. –

Not that I doubt a long course of "competitive highness" will *ultimately make the organisation higher* in every sense of the word; but it seems most difficult to test it. Look at the Erigeron Canadensis on one hand, and Anacharis on the other; these plants must have some advantage over European productions to spread as they have; yet who could discover it? Monkeys can co-exist with Sloths and Opossums, orders at the bottom of scale; and the opossums might well be beaten by placental insectivores coming from a country, where there were no monkeys &c, &c. – I should be sorry to be forced to give up the view that an old and very large continuous territory would generally produce organisms higher in the competitive sense than a smaller territory. I may of course be quite wrong about plants of Australia (and your facts are of course quite new to me on their highness) but when I read the accounts of immense spreading of European plants in Australia, and think of the wool and corn brought thence to Europe, and not one plant naturalised, I can hardly avoid suspicion that Europe beats Australia in its productions. If many (i.e. if more than 1 or 2) Australian plants are *truly* naturalised in India (N.B Naturalisation on Indian mountains hardly *quite* fair, as mountains are small islands in the land) I must strike my colours. I should be glad to hear whether what I have written very obscurely on this point produces *any* effect on you: for I want to clear my mind, as perhaps I should put a sentence or two in my abstract [the *Origin!*] on this subject.

(*Darwin, 1985, 7, 228–9; letter from Darwin to Hooker, December 31, 1858*)

For all his worries, however, Darwin was committed to the idea of progress, he thought it came through natural selection, and it led to ever-greater complexity, and ended (at least thus far) with humankind. He said this repeatedly, most famously at the end of the *Origin of Species.*

It is interesting to contemplate an entangled bank, clothed with many plants of many kinds, with birds singing on the bushes, with various insects flitting about, and with worms crawling through the damp earth, and to reflect that these elaborately constructed forms, so different from each other, and dependent on each other in so complex a manner, have all been produced by laws acting around us. These laws, taken in the largest sense, being growth with reproduction; inheritance which is almost implied by reproduction; variability from the indirect and direct action of the external conditions of life, and from use and disuse; a ratio of increase so high as to lead to a struggle for life, and as a consequence to Natural Selection, entailing Divergence of Character

and the Extinction of less-improved forms. Thus, from the war of
nature, from famine and death, the most exalted object which we
are capable of conceiving, namely, the production of the higher
animals, directly follows. There is grandeur in this view of life,
with its several powers, having been originally breathed into a few
forms or into one; and that, whilst this planet has gone cycling on
according to the fixed law of gravity, from so simple a beginning
endless forms most beautiful and most wonderful have been, and
are being, evolved.

> (Darwin, 1859, 489–490)

From natural selection, the "higher animals" follow directly.
From "so simple a beginning", we get beautiful and wonderful forms.
Clearly, in some sense, in the Darwinian world picture, this comes
about because those that win in the struggle overall are smarter, more
gifted, more complex than the losers. At least in the animal world this
seems to be so. Darwin's good friend, the American botanist Asa Gray,
pointed out that things are a lot murkier in the plant world and it is
by no means obvious what would constitute "higher" in such cases.

In reply to your question –

If Oak and Beech had large-colored corolla, &c – I know of no reason why it would be reckoned a low form, but the contrary, quite.
But we have no basis for high & low in any class – say Dicotyle-
dons, except perfection of development or the contrary in the flo-
ral organs, – and even the envelopes, – and as we know these may
be reduced to any degree in any order or group, we have really that
I know of, no philosophical basis for high & low. Moreover, the
vegetable kingdom does not *culminate*, as the *animal* kingdom
does. It is not a *kingdom*, but a commonwealth – a *democracy*,
and therefore puzzling and unaccountable from the former point of
view.

> (Darwin, 1985, 11, 91; letter from Asa Gray to Darwin, January 23,
> 1863)

There is no absolute necessity for upwards change. Darwin was always clear on that. He wanted to put a firm line between himself and Germanic-type notions of inevitable upwards progress. For this reason, although he certainly subscribed to some kind of analogy between embryological development and the history of life – "ontogeny recapitulates phylogeny", in the famous declaration of the German biologist Ernst Haeckel – I don't think he ever saw the two as that closely entwined, with the nigh-inevitability of development of the embryo being mirrored by a nigh-inevitable development in the history of life overall. It was just that more and more sophisticated organisms did evolve, and overall they would have an edge. But what was the criterion of sophistication, of complexity as it were? Darwin was influenced in his thinking here both by the work of the leading German embryologists – Karl Ernst von Baer in particular – and by the metaphor of a division of labor, always an attractive notion in his thinking. No doubt at this point he was influenced by his family background. The Darwin–Wedgwood family made a huge amount of money out of the Industrial Revolution, and the division of labor was perhaps the key guiding principle at work here.

In the third edition of the *Origin* (published in 1861), Darwin wrote as follows:

> *On the degree to which Organisation tends to advance.* – Natural selection acts, as we have seen, exclusively by the preservation and accumulation of variations, which are beneficial under the organic and inorganic conditions of life to which each creature is at each successive period exposed. The ultimate result will be that each creature will tend to become more and more improved in relation to its conditions of life. This improvement will, I think, inevitably lead to the gradual advancement of the organisation of the greater number of living beings throughout the world. But here we enter on a very intricate subject, for naturalists have not defined to each other's satisfaction what is meant by an advance in organisation. Amongst the vertebrata the degree of intellect and

an approach in structure to man clearly come into play. It might
be thought that the amount of change which the various parts and
organs undergo in their development from the embryo to maturity would suffice as a standard of comparison; but there are cases,
as with certain parasitic crustaceans, in which several parts of
the structure become less perfect, so that the mature animal cannot be called higher than its larva. Von Baer's standard seems the
most widely applicable and the best, namely, the amount of differentiation of the different parts (in the adult state, as I should
be inclined to add) and their specialisation for different functions;
or, as Milne Edwards would express it, the completeness of the
division of physiological labour.

(Darwin, 1861, 133)

He then went on immediately to note that this doesn't seem to work
for all animals and even less so for plants. But he did think that he
was onto something – and (stressing that the brain is much involved)
once again showed that that something included humans!

If we look at the differentiation and specialisation of the several organs of each being when adult (and this will include the
advancement of the brain for intellectual purposes) as the best
standard of highness of organisation, natural selection clearly
leads towards highness; for all physiologists admit that the specialisation of organs, inasmuch as they perform in this state
their functions better, is an advantage to each being; and hence
the accumulation of variations tending towards specialisation is
within the scope of natural selection.

(Darwin, 1861, 134)

Note that this is at the level of the individual organism. You
might think that there are levels of complexity, from the individual to
the group and perhaps up to what we might now call the eco-system,
and even eventually the whole of life. As forms proliferate and interact, we get ever more complex systems. Darwin was certainly fully
aware of this, however there had to be (in his thinking) a fundamental

difference between any complexity at the individual level and any complexity at the group level. For Darwin, natural selection always worked at and for the individual – where admittedly the individual could comprise the interrelated nest of the social insect (Ruse, 1980). Natural selection could never work for the species or for the higher group or system. If complexity did come at this level it had to be a byproduct of selection working for the individual. And this is what he thought he had covered in his celebrated "principle of divergence". We get a whiff of this in his early notebooks that he kept in the late 1830s when he was formulating his theory.

> The enormous *number* of animals in the world depends of their varied structure & complexity. – hence as the forms became complicated, they opened *fresh* means of adding to their complexity. – but yet there is no *necessary* tendency in the simple animals to become complicated although all perhaps will have done so from the new relations caused by the advancing complexity of others. – It may be said, why should there not be at any time as many species tending to dis-development (some probably always have done so, as the simplest fish), my answer is because, if we begin with the simplest forms & suppose them to have changed, their very changes tend to give rise to others.
>
> (Barrett et al., 1987, Notebook, E, 95–6)

I will come back to this passage in a moment, but now move on to the *Origin* and the full statement of the Principle of Divergence. Darwin starts the discussion, as he starts many discussions in the *Origin*, by referring to the domestic world and about how breeders have different ends and select for those ends. A sportsman wants a fleet racehorse and a farmer wants a sturdy draught animal and so we get a divergence of form. Then onto the natural world.

> But how, it may be asked, can any analogous principle apply in nature? I believe it can and does apply most efficiently, from the simple circumstance that the more diversified the descendants from any one species become in structure, constitution, and

habits, by so much will they be better enabled to seize on many
and widely diversified places in the polity of nature, and so be
enabled to increase in numbers.

(Darwin, 1859, 112)

As I have said, this is a matter of benefits to the individual not the
group.

> The truth of the principle, that the greatest amount of life can be
> supported by great diversification of structure, is seen under many
> natural circumstances. In an extremely small area, especially if
> freely open to immigration, and where the contest between indi-
> vidual and individual must be severe, we always find great diver-
> sity in its inhabitants.

(Darwin, 1859, 114)

Selection is not for the "greatest amount of life." This comes as a consequence of the individual contest. Having said this, the division of
labor applies here at the group level as much as at the individual level.

> The advantage of diversification in the inhabitants of the same
> region is, in fact, the same as that of the physiological division
> of labour in the organs of the same individual body – a subject so
> well elucidated by Milne Edwards. No physiologist doubts that
> a stomach by being adapted to digest vegetable matter alone, or
> flesh alone, draws most nutriment from these substances. So in
> the general economy of any land, the more widely and perfectly
> the animals and plants are diversified for different habits of life, so
> will a greater number of individuals be capable of their support-
> ing themselves. A set of animals, with their organisation but little
> diversified, could hardly compete with a set more perfectly diver-
> sified in structure.

(Darwin, 1859, 115–116)

It is not the sets as such that are competing but the individuals within
the sets. Of course, you might want to say – I presume that this is
the exercise of showing how you would want to say – that there is

nevertheless complexity at the set or group level. It is just that, in Darwin's case, this would be a byproduct of individual interests – and the individual interests are certainly not complexity at the set level, as such. This rather points to the fact that Darwin does allow that some notions of complexity, although perhaps ultimately relatable back to selection, do not themselves emerge as the direct product of selection. And now, going back to the passage quoted from the notebook, perhaps – although this may only have been true of the early years – in a deeper sense Darwin allowed a kind of complexity to emerge without the direct aid of selection. The first part of the passage does seem to call out for selection at work and to be a forerunner of the later Principle of Divergence. "The enormous *number* of animals in the world depends of their varied structure & complexity. – hence as the forms became complicated, they opened *fresh* means of adding to their complexity." (If this interpretation is indeed correct, then as a side point we have somewhat of a corrective to Darwin's late-life memory that he did not get the Principle until about 1850 – he was certainly onto something before that.) The second part of the passage, however, seems to suggest something along the lines of complexity emerging rather as a consequence of the nature of things – most particularly that you can go to complexity from simplicity simply by random forces. "It may be said, why should there not be at any time as many species tending to dis-development (some probably always have done so, as the simplest fish), my answer is because, if we begin with the simplest forms & suppose them to have changed, their very changes tend to give rise to others." It may just be that at this point Darwin was not really fully aware of what he could do with selection and that later he dropped this speculation. Either way, note that Darwin was insistent that there is no necessity at work. We do not have a kind of Hegelian world spirit pushing us ever upwards.

12.2 HERBERT SPENCER

Move across now to Darwin's contemporary Herbert Spencer. A fellow Englishman, Spencer wrote on topics across the spectrum – but

at the heart was his own rather idiosyncratic view of the evolutionary process. Like Darwin, Spencer's thinking on evolution was much
influenced – perhaps "inspired" is a better word – by the thinking
of the political economist Thomas Robert Malthus. Like Darwin,
Spencer (1852) hit independently on the notion of natural selection.
(He did this after Darwin found the notion but published before Darwin published.) And like Darwin, he came up with a name for the
process – one which many preferred to "natural selection" and which
Darwin himself introduced into later editions of the *Origin* – "the
survival of the fittest".

However, their views of the evolutionary process could not have
truly been more different. Selection was always a minor phenomenon
for Spencer, a process that may have done a bit of cleaning up after
the main work was done. For Spencer the main force of evolutionary change, down at the biological level, was so-called Lamarckism –
the inheritance of acquired characteristics. Whereas Darwin saw the
Malthusian pressures from population growth as leading to a kind
of selection – only some are going to get through – Spencer saw the
Malthusian pressures as stimulating organisms to do better and, as
this happens, the results get incorporated into the new generations
automatically and change eventuates. But what kind of change? Here
Spencer drew on some very Victorian beliefs about organisms having only so much vital bodily fluid and as it is used in one way, for
instance reproduction, so it is not available to be used in other ways,
for instance in building brains and conferring intelligence. Under the
Malthusian pressures, brains and intelligence do emerge and at the
same time the reproductive abilities and habits start to drop away –
herrings to humans – and so eventually the population pressures
decline and basically go away (Ruse, 1999).

This isn't just a matter of things getting better. This is the
Platonic form of progress, a point that Spencer confirmed a year or
two before the *Origin* was published. In his 1857 essay, "Progress:
its law and cause", Spencer staked his banner: "Now, we propose
in the first place to show, that this law of organic progress is the

law of all progress. Whether it be in the development of the Earth, in the development of Life upon its surface, in the development of Society, of Government, of Manufactures, of Commerce, of Language, Literature, Science, Art, this same evolution of the simple into the complex, through successive differentiations, holds throughout."

How did Spencer characterize the complex? In respects, he drew on the same ideas as Darwin.

> Higher organisms are distinguished from lower ones partly by bulk, and partly by complexity. This complexity essentially consists in the mutual dependence of numerous different organs, each subserving the lives of the rest, and each living by the help of the rest. Instead of being made up of many like parts, performing like functions, as the Crinoid, the Star-fish, or the Millipede, a vertebrate animal is made up of many unlike parts, performing unlike functions. From that initial form of a compound organism, in which a number of minor individuals are simply grouped together, we may, more or less distinctly, trace not only the increasing closeness of their union, and the gradual disappearance of their individualities in that of the mass, but the gradual assumption by them of special duties. And this "physiological division of labour", as it has been termed, has the same effect as the division of labour amongst men. As the preservation of a number of persons is better secured when, uniting into a society, they severally undertake different kinds of work, than when they are separate and each performs for himself every kind of work; so the preservation of a congeries of parts, which, combining into one organism, respectively assume nutrition, respiration, circulation, locomotion, as separate functions, is better secured than when those parts are independent, and each fulfils for itself all these functions.
>
> *(Spencer, 1852, 486)*

To characterize the phenomenon of complexity, Spencer introduced two terms which have had a very long shelf life. Simple organisms (or states of affairs generally) are "homogeneous".

Complex organisms (or states) are "heterogeneous". The course of progress therefore is from the homogeneous to the heterogeneous. (As it happens, although Spencer was never very generous in acknowledging influences, he got this idea from the German philosopher Friedrich Schelling, via Schelling's English interpreter – some would say "plagiarist" – the poet William Coleridge (Ruse, 2013b).) As he developed his thinking around the time that Darwin published the *Origin*, Spencer fell more and more under the spell of physics. This led to his full-blown theory of "dynamic equilibrium". He argued that organisms, or a state, exist in a balance – working normally, with everything in equilibria. Then something – quite possibly something from outside – acts to upset the balance and things get into play. They are striving always to regain equilibrium, just as a ball bearing at the bottom of a bowl, if disturbed, will naturally tend back down to the center and to regained equilibrium. Eventually this balance will be achieved, but it will be a new balance, higher than before, and with a greater heterogeneity than before. Backing all of this, Spencer gave a kind of metaphysical principle – one that may or may not have been behind the notebook passage of Darwin quoted above – that causes always proliferate in several effects but never conversely. So complexity emerges naturally. "Every active force produces more than one change – every cause produces more than one effect" (Spencer, 1857, 32).

There was overlap between Darwin and Spencer, but essentially we have two world visions, two paradigms if you like. On the one hand, we have Darwin's view where processes are driven by selection, where the end result has to be adaptive (of value to the individual), where progress is expected but not guaranteed, and where intelligence is the likely outcome (a point made very clear in the *Descent of Man*). On the other hand, we have Spencer's view where processes (especially in the later versions of his thinking) are driven more by the metaphysical principles that make matter work as it does, where increasingly adaptation is not the crucial question, where progress really is pretty much guaranteed (given enough time), and where intelligence is simply bound to appear. What I would suggest and will try to show – albeit very briefly – is that this is a division that persisted after the main players left the stage, has had a shelf life right down through the one hundred and fifty years since the ideas were being formulated, and can be found even unto this day.

12.3 FISHER AND WRIGHT

Let us skip forward now to about 1930. After the *Origin* was published, it is very well known that, although the idea of evolution was taken up rapidly (with exceptions of places like the American South where it is still not widely accepted), the mechanism of natural selection found little acceptance. It was not until around 1900 that thinking about heredity was put on a firm basis (thanks to the rediscovery of Mendel's work), and biologists could now start to see how selection could indeed operate and be a significant force for change (Provine, 1971). By about 1930, mathematically gifted thinkers were working on the problems of change in populations and finally a proper foundation for evolutionary thought was in play. Ronald A. Fisher in England and Sewall Wright in America were the two major thinkers responsible for the development of (what was known in England as) neo-Darwinism or (what was known in America as) the Synthetic Theory of Evolution. (There was a third significant thinker, J. B. S. Haldane in Britain, but at the time he had no students or school to carry on his work and line of thought.)

My claim is that you see in Fisher's thinking a continuation of Darwin's thinking and in Wright's thinking a continuation of Spencer's thinking. Start with Fisher and with his 1930 masterwork, *The Genetical Theory of Natural Selection*. The background is that Fisher was a fanatical Darwinian, emotionally as well as intellectually. (For many years, he was aided financially by Major Leonard Darwin, the youngest child of Darwin himself.) He was also (unlike Darwin) a deeply committed Christian (he was an Anglican), who thought that Darwinian evolution was God's way of creation and that it had led up to humankind (Ruse, 1996). He feared that now humans are ignoring evolution and that this is leading to decline – in intelligence basically – and that it is our Christian moral duty to reverse this trend. This he hoped to do through eugenics, particularly the positive eugenics of getting the more worthwhile and talented to breed more than they do. (The reason why Fisher needed financial handouts was that he put his beliefs into practice and married a girl of good breeding stock and had a very large family. The reason why Leonard Darwin was so supportive was that he was head of the British eugenics society.)

How does progress take place? We are now in the Mendelian era, so for Fisher (and for the other "population geneticists") the starting point – by this I mean the conceptual starting point, the premises from which deductions will flow – was the generalization of Mendelian genetics to groups, the so-called Hardy Weinberg law. This is basically an equilibrium law and functions much like Newton's first law of motion, which is to say that if nothing happens then nothing happens. In the case of biology, gene ratios (the proportions of the units of heredity) will stay the same unless there are forces disrupting them. Now forces are introduced, one of the chief of which is natural selection. This will change gene ratios, which translated into the physical world means a change in the nature of organisms – in other words, evolution.

Fisher, as a student at Cambridge, had taken a course with the gas theorist James Jeans, and this always colored his thinking about the nature of evolutionary change. He saw populations of organisms as clouds of genes randomly mixing together. Thanks to the Hardy Weinberg law, they will stay as they are until something acts on them. Selection now enters in and starts to bring on change. But how does this happen. It is at this point that Fisher introduced his somewhat mysterious "Fundamental Theorem of Evolution":

> The rate of increase in fitness of any organism at any time is equal to its genetic variance in fitness at that time.
>
> (Fisher, 1930, 35)

I speak of it as "mysterious" because at the time no one seemed to know what to make of it and, candidly, few seem to know to this day what to make of it. It is surely significant that Fisher's great supporter and collaborator, E. B. Ford (1931), quite ignored it in the little popular book he wrote just after the *Genetical Theory* appeared! But basically it is saying that if you have variation in a population (variation making some better than others), then there is going to be movement towards the better or best. (And the rate of change is going to be a function of how big the differences are.) In other words, in your clouds of genes, you are going to have selection-fueled progress up to the best ones. Fisher of course was aware that all sorts of other factors could get involved – mutation, migration, general change of external circumstances, and so forth. So you are not going to get steady progress to the heights. It is just that there is a general tendency that way and ultimately the best, namely humans, emerged. My suspicion is that Fisher's Christian beliefs led him to see more guarantees than Darwin felt truly justified, but this is a hunch.

Sewall Wright had a very different picture of things (Provine, 1986). According to his "shifting balance theory of evolution", organisms in a population (or a bigger group) are distributed across an "adaptive landscape" (Wright, 1931, 1932). Generally, they rest on or close to the peaks. (So Wright is certainly not indifferent to selection. You are on the top of a peak because that is the place that is most adaptively advantageous, and selection put you there. But read on.) Now the question is how things go from there, and how organisms might come off one peak and scale another – and if the newly scaled peak is going to be higher in some sense than the old one (in other words, is there going to be some kind of overall progress)? It is here that Wright became really innovative. Drawing on ten years' experience of working at the US Department of Agriculture, Wright believed that the maximum effect would happen if groups were fragmented, and if change occurred first in the fragments and then, on recombination, the best changes could spread through the whole group. It was at this point Wright made his profoundly non-Darwinian move, because he was able to show that in small groups the vagaries, the randomness, of breeding encounters could outpower selection – non-Darwinian "genetic drift" could be the cause of change. Moreover, he thought that this is where really creative new innovations could and would appear. So the fuel of change, as it were, consists of new innovations brought on by genetic drift, which are then reintroduced into the general population (or the general population is reconstituted) and now, thanks to selection, a new peak is scaled.

What was the inspiration behind this essentially non-Darwinian theory of change? We know that Wright was deeply influenced as a student by Spencer enthusiasts, including his own father and then, when a graduate student at Harvard, by the chemist L. J. Henderson (Ruse, 2004). But there is no need to rely on origins. The evidence is right there on the surface. Most particularly, the language flags you. See what Wright himself says about his theory.

> Evolution as a process of cumulative change depends on a proper balance of the conditions, which, at each level of organization – gene, chromosome, cell individual, local race – make for genetic homogeneity or genetic heterogeneity of the species . . . The type and rate of evolution in such a system depend on the balance among the evolutionary pressures considered here.
>
> (Wright, 1931, 158)

The Spencerian language is right there – homogeneity or heterogeneity. Moreover, something acts (probably from outside) to disturb this picture, breaking the groups up into fragments. The random forces of nature now take over, creating new and better. Then everything comes back together again and equilibrium is regained. But why is it necessarily better? In a sense, Wright, like Spencer, takes this more or less as a given. The landscape could be like a waterbed, where there is no progress because as one part goes up another goes down. But Wright obviously thinks of it more as something made from rock, from granite, and basically there are always ever higher peaks to be climbed. Wright, incidentally, did not put this all in a Christian perspective as did Fisher, but there were some powerful underlying metaphysical urges, including a belief in panpsychic monism – that everything is alive and conscious and the world is moving towards a universal mind of some sort.

12.4 EVOLUTIONARY THOUGHT TODAY

Start with the Darwinians, and take Richard Dawkins as an exemplar. As far as characterizing complexity, sometimes he is rather wary. He is scornful of attempts to pin down quantities of progress, in any absolute sense. He rips into such notions concluding: "I recommend that evolutionary writers should no longer, under any circumstances, use the adjectives 'higher' and 'lower'" (Dawkins, 1992, 272). At other times, however, he is more optimistic. Starting with ideas in information theory, he thinks that more complex organisms would require physically longer descriptions than less complex organisms.

> We have an intuitive sense that a lobster, say, is more complex (more "advanced", some might even say more "highly evolved") than another animal, perhaps a millipede. Can we *measure* something in order to confirm or deny our intuition? Without literally turning it into bits, we can make an approximate estimate of the information contents of the two bodies as follows. Imagine writing the book describing the lobster. Now write another book describing the millipede down to the same level of detail. Divide the word-count in the one book by the word-count in the other, and you will have an approximate estimate of the relative information content of lobster and millipede. It is important to specify that both books describe their respective animals "down to the same level of detail". Obviously, if we describe the millipede down to cellular detail, but stick to gross anatomical features in the case of the lobster, the millipede would come out ahead.
>
> But if we do the test fairly, I'll bet the lobster book would come out longer than the millipede book.
>
> (Dawkins, 2003, 100)

Let us skate quickly over the troublesome question of what constitutes "doing the test fairly", not mention problematic issues about whether sheer size or number is always that helpful a marker – genome size is certainly often a bad guide to what we might intuitively think of as higher or lower (Gregory, 2001). Do we get progress, in the sense of greater complexity? Dawkins thinks we do and, like Darwin, he is keen to tie it in with adaptive advantage. Dawkins (1997, 1016) gave the following definition: "A tendency for lineages to improve cumulatively their adaptive fit to their particular way of life, by increasing the numbers of features which combine together in adaptive complexes." You might think this a little bit wishy washy, because the characterization seems to say little or nothing about that all-important creature, *Homo sapiens*. But elsewhere Dawkins speaks to that issue. In his great popular overview of modern evolutionary thinking, *The Blind Watchmaker* (1986), Dawkins refers to Harry Jerison's (1973) notion of an Encephalization Quotient, this being a kind of universal animal IQ, that works from brain size and subtracts the gray matter simply needed to get the body functioning – whales necessarily have bigger brains than shrews, because they have bigger bodies. What counts is what is left when you take off the body-functioning portion. Thus measured, humans come way out on top, leading Dawkins (1986, 189) to reflect: "The fact that humans have an EQ of 7 and hippos an EQ of 0.3 may not literally mean that humans are 23 times as clever as hippos!" But, he concludes, it does tell us "something".

Dawkins (1989) has also tied in his thinking about progress with the notion of the "evolution of evolvability". Sometimes, you just get evolutionary breakthroughs – like the eukaryotic cell – that have more potential, and hence evolution has made a jump to a new dimension.

Notwithstanding Gould's just skepticism over the tendency to label each era by its newest arrivals, there really is a good possibility that major innovations in embryological technique open up new vistas of evolutionary possibility and that these constitute genuinely progressive improvements (Dawkins, 1989; Maynard Smith & Szathmáry, 1995). The origin of the chromosome, the bounded cell, organized meiosis, diploidy and sex, the eukaryotic cell, multicellularity, gastrulation, molluscan torsion, segmentation – each of these may have constituted a watershed event in the history of life. Not just in the normal Darwinian sense of assisting individuals to survive and reproduce, but watershed in the sense of boosting evolution itself in ways that seem entitled to the label progressive. It may well be that after, say, the invention of multicellularity, or the invention of metamerism, evolution was never the same again. In this sense, there may be a one-way ratchet of progressive innovation in evolution.

(Dawkins, 1997, 1019–1020)

Dawkins has always made brilliant use of metaphor – selfish gene, blind watchmaker, mount improbable – and metaphor is much involved in the thinking about progress. In *The Blind Watchmaker*, the metaphor of bigger and bigger on-board computers (aka brains) plays a vital role, as it has elsewhere.

Computer evolution in human technology is enormously rapid and unmistakably progressive. It comes about through at least partly a kind of hardware/software coevolution. Advances in hardware are in step with advances in software. There is also software/software coevolution. Advances in software make possible not only improvements in short-term computational efficiency – although they certainly do that – they also make possible further advances in the evolution of the software. So the first point is just the sheer adaptedness of the advances of software make for efficient computing. The second point is the progressive thing. The advances of software open the door – again I wouldn't mind using the word "floodgates" in some instances – open the floodgates to further advances in software.

(Ruse, 1996, 469. This is from a presentation given in Melbu, Norway, in 1989)

Evolution is cumulative, for it has "the power to build new progress on the shoulders of earlier generations of progress". And brains, especially the biggest and best brains, are right there at the heart, or (perhaps we should say) end: "I was trying to suggest by my analogy with software/software coevolution, in brain evolution that these may have been advances that will come under the heading of the evolution of evolvability in [the] evolution of intelligence" (Ruse, 1996, 469, quoting Dawkins in Melbu).

But what is the causal force behind all of this, or rather, since we are working in a Darwinian mode, how does natural selection work to bring all of this about? In much the way that Darwin himself supposed – organisms beat out their predecessors and thus climb upwards. There is a bit of a twist on this, namely that much of the improving change is seen as coming from competition between lines – "arms races". In the little book on orchids that Darwin wrote just after the *Origin*, he actually talked about "races" between plants and the insects that use them, a point picked up (with some wariness) by Asa Gray. "Of course we believers in real design make the most of your "frank" and natural terms, "contrivance, purpose," etc., and pooh-pooh your endeavors to resolve such contrivances into necessary results of certain physical processes, and make fun of the race between long noses and long nectarines" (Gray, 1894, 2, 502). The person who really developed the notion in the military sense, transferring it across to biology, was Thomas Henry Huxley's grandson, Julian Huxley. Right from his first little book, *The Individual in the Animal Kingdom*, written before the First World War, Huxley was likening biological evolution to the competition between nations in preparation for war. Germany and Britain were competing on the sea, leading Huxley to write: "The leaden plum-puddings were not unfairly matched against the wooden walls of Nelson's day". He then added that today "though our guns can hurl a third of a ton of sharp-nosed steel with dynamite entrails for a dozen miles, yet they are confronted with twelve-inch armor of backed and hardened steel, water-tight compartments, and targets moving thirty miles an hour. Each advance in attack has brought forth, as if by magic, a corresponding advance in defence." Explicitly, Huxley likened this to the organic world, for "if one species happens to vary in the direction of greater independence, the inter-related equilibrium is upset, and cannot be restored until a number of competing species have either given way to the increased pressure and become extinct, or else have answered pressure with pressure, and kept the first species in its place by themselves too discovering means of adding to their independence" (Huxley, 1912, 115–16). And so finally: "it comes to pass that the continuous change which is passing through the organic world appears as a succession of phases of equilibrium, each one on a higher average plane of independence than the one before, and each inevitably calling up and giving place to one still higher".

Dawkins picks right up on this sort of thinking. Organisms compete against each other – more precisely, organisms of one group compete against the organisms of another group – and the interaction brings adaptive changes to both sides. Classically, the prey runs a little faster, and then the predator has to run a little faster – or starve. We have something akin to the human notion of an arms race, and as with the human notion, we get improvement. Of course you might object that, in a way, this is all relative progress – does one really want to say an efficient gun is absolutely good? – and Dawkins much to his credit exploits this idea to the scientific full. Alone and with John Krebs he has offered careful and fruitful analysis of the ways in which arms races can and might be expected to function (Dawkins & Krebs, 1979). For instance, he distinguishes between asymmetrical arms races (with different kinds of competitors, like prey and predator) and symmetrical arms races (with similar competitors, as one might get in sexual selection). However, what about some kind of absolute progress – specifically what about getting humans? Does one get these from arms races? My suspicion is that Dawkins rather thinks that in the end one does. In the paper co-authored with Krebs he writes that even "if modern predators are no better at catching prey than Eocene predators were at catching Eocene prey, it does at first sight seem to be an expectation of the arms race idea that modern predators might massacre Eocene prey. And Eocene predators chasing modern prey might be in the same position as a Spitfire chasing a jet" (Dawkins & Krebs, 1979, 490). In other words, in Dawkins's happy Darwinian world, it all comes out right in the end.

Do we still today have a Spencerian alternative? There are certainly those who think we can get some kind of complexity, in a progressive sense, without natural selection; people who are at least in the Spencerian tradition of thinking that the laws of nature themselves can give you all that you need. The most prominent today is the theoretical biologist Stuart Kauffman. He writes: "The tapestry of life is richer than we have imagined. It is a tapestry with threads of accidental gold, mined quixotically by the random whimsy of quantum events acting on bits of nucleotides and crafted by selection sifting. But the tapestry has an overall design, an architecture, a woven cadence and rhythm that reflects underlying law – principles of self organization" (Kauffman, 1995, 185). Kauffman is not alone in thinking like this. Before him were people like the Scottish morphologist D'Arcy Wentworth Thompson (1917), who thought that the laws of form could produce all of the complexity that you find or need in this world. Along with Kauffman, perhaps at a slightly more ethereal level (I do not mean that as a criticism), was the late Canadian-born, England-residing biologist Brian Goodwin. Defining a field as "the behavior of a dynamic system that is extended in space," Goodwin writes:

A new dimension to fields is emerging from the study of chemical systems such as the Beloussov–Zhabotinsky reaction [a chemical process that produces intricate patterns akin to those found in living organisms] and the similarity of its spatial patterns to those of living systems. This is the emphasis on self-organization, the capacity of these fields to generate patterns spontaneously without any specific instructions telling them what to do, as in a genetic program. These systems produce something out of nothing... There is no plan, no blueprint, no instructions about the pattern that emerges. What exists in the field is a set of relationships among the components of the system such that the dynamically stable state into which it goes naturally – what mathematicians call the generic (typical) state of the field – has spatial and temporal pattern.

(Goodwin, 2001, 51)

With people like Kauffman, Thompson, and Goodwin, it is not always easy to see if they are arguing that nature alone can produce complexity that is adaptive, or if they are arguing that adaptation is simply not that important. Clearly, in a sense we have organisms doing what they need to do to survive and reproduce. The point is that what brings this about is not a focus on needs and ends, but rather the unfurling of things according to the internal forces of nature – self organization. To use a nice phrase by Stuart Kauffman, we have "order for free".

Getting closer to Spencer himself perhaps was Stephen Jay Gould – perhaps no great surprise from someone who called his great theory "punctuated equilibrium". In his book *Full House* (1996), he argued that there is certainly no necessary, selection-driven process of progressive evolution leading to greater complexity. However, he did at the same time agree that the natural course of events would lead to ever-more complex organisms. His point was that there had to be a one-way process. There is a limit on the simplicity of organisms. Below a certain point, they can get no more simple. However, there is no limit on the potential complexity of organisms. It is all rather like a drunkard walking along a sidewalk, with the road on one side and a wall on the other. Eventually, through random processes, the drunkard will fall into the gutter. He cannot go further the other way, because the wall prevents this.

These ideas have been picked up and developed at some length recently in a new book by two academics at Duke University, one a biologist (Daniel McShea) and the other a philosopher (Robert Brandon). In *Biology's First Law: the Tendency for Diversity and*

Complexity to Increase in Evolutionary Systems, they offer a very Spencerian vision of the evolutionary process. They call their law the "zero-force evolutionary law", or ZFEL. It is formulated as follows: "In any evolutionary system in which there is variation and heredity, in the absence of natural selection, other forces, and constraints acting on diversity or complexity, diversity and complexity will increase on average". What do McShea and Brandon mean by complexity? Certainly nothing to do with natural selection as such. They say that complexity is "a function only of the amount of differentiation among parts within an individual". Elsewhere they say "'complexity' just means number of parts types or degree of differentiation among parts". They are very careful to specify that this has nothing to do with adaptation. Indeed they say "in our usage, even functionless, useless, part types contribute to complexity. Even maladaptive differentiation is pure complexity". How could this complexity come about? It all seems to be a matter of randomness. With Gould, and I think with Spencer, they simply believe that over time more and more things will happen and pieces will be produced and thus complexity will emerge. It is the inevitability of the drunkard falling into the gutter.

12.5 HOW DO WE CHOOSE?

In this chapter I have been approaching matters from a historical perspective. But ultimately I am interested in the philosophical juice that one might extract. My strong sense is that the Darwinians and the Spencerians are talking past each other. They have different paradigms, if you like. Darwinians think that unless one tries to attach adaptation to complexity, then the effort is not worthwhile. Spencerians beg to differ, arguing that complexity is a perfectly good notion in its own right, whether or not adaptation is involved at all. I'm not quite sure how one would resolve this issue. My suspicion is that Darwinians would argue that their theory overall is simply better when it comes to understanding the world. They can combine things under one explanation, make predictions, and do much more that one expects of good theories. Obviously Spencerians will counter that this is not so. They argue (as people like Gould certainly argued) that there is much more to the organic world than adaptation. If one does not take account of this, one is distorting the true world picture. But how does one move beyond this dispute? There are times when I want to urge that the whole topic of complexity, organization, progress be dropped. Darwinians have the right theory but they cannot handle the terms. Spencerians can handle the terms but they have the wrong theory. And yet! Was Darwin so very wrong in the sentiment, taken from the *Origin of Species*, expressed right at the beginning of this chapter? That I'm afraid I'm going to have to leave as an exercise for the reader!

REFERENCES

Barrett, P. H., Gautrey, P. J., Herbert, S., Kohn, D., & Smith, S. (eds.) (1987). *Charles Darwin's Notebooks, 1836–1844*. Ithaca, NY: Cornell University Press.

Darwin, C. (1859). *On the Origin of Species by Means of Natural Selection, or the Preservation of Favoured Races in the Struggle for Life*. London: John Murray.

Darwin, C. (1861). *Origin of Species* (Third Edition). London: John Murray.

Darwin, C. (1985). *The Correspondence of Charles Darwin*. Cambridge: Cambridge University Press.

Dawkins, R. (1986). *The Blind Watchmaker*. New York, NY: Norton.

Dawkins, R. (1989). The evolution of evolvability. In C. G. Langton (ed.), *Artificial Life*. 201–20. Redwood City, CA.: Addison-Wesley.

Dawkins, R. (1992). *Progress*. In E. F. Keller & E. Lloyd (eds.), *Keywords in Evolutionary Biology*. 263–272. Cambridge, MA: Harvard University Press.

Dawkins, R. (1997). Human chauvinism: review of *Full House* by Stephen Jay Gould. *Evolution* **51**, 3, 1015–1020.

Dawkins, R. (2003). *A Devil's Chaplain: Reflections on Hope, Lies, Science and Love*. Boston and New York: Houghton Mifflin.

Dawkins, R. & Krebs, J. R. (1979). Arms races between and within species. *Proceedings of the Royal Society of London, B*, **205**, 489–511.

Fisher, R. A. (1930). *The Genetical Theory of Natural Selection*. Oxford: Oxford University Press.

Ford, E. B. (1931). *Mendelism and Evolution*. London: Methuen.

Goodwin, B. (2001). *How the Leopard Changed its Spots*, Second Edition. Princeton: Princeton University Press.

Gould, S. J. (1996). *Full House: the Spread of Excellence from Plato to Darwin.* New York, NY: Paragon.

Gray, J. L. (1894). *Letters of Asa Gray.* Boston: Houghton, Mifflin.

Gregory, T. R. (2001). Coincidence, coevolution, or causation? DNA content, cell size, and the C-value enigma. *Biological Reviews, 76,* 65–101.

Huxley, J. S. (1912). *The Individual in the Animal Kingdom.* Cambridge: Cambridge University Press.

Jerison, H. (1973). *Evolution of the Brain and Intelligence.* New York, NY: Academic Press.

Kauffman, S. A. (1995). *At Home in the Universe: the Search for the Laws of Self-Organization and Complexity.* New York: Oxford University Press.

McShea, D. & Brandon, R. (2010). *Biology's First Law: the Tendency for Diversity and Complexity to Increase in Evolutionary Systems.* Chicago: University of Chicago Press.

Maynard Smith, J. & Szathmary, E. (1995). *The Major Transitions in Evolution.* New York: Oxford University Press.

Provine, W. B. (1971). *The Origins of Theoretical Population Genetics.* Chicago: University of Chicago Press.

Provine, W. B. (1986). *Sewall Wright and Evolutionary Biology.* Chicago: University of Chicago Press.

Richards, R. J. (1987). *Darwin and the Emergence of Evolutionary Theories of Mind and Behavior.* Chicago: University of Chicago Press.

Ruse, M. (1980). Charles Darwin and group selection. *Annals of Science, 37,* 615–630.

Ruse, M. (1996). *Monad to Man: the Concept of Progress in Evolutionary Biology.* Cambridge, MA: Harvard University Press.

Ruse, M. (1999). *The Darwinian Revolution: Science Red in Tooth and Claw.* Second edn. Chicago: University of Chicago Press.

Ruse, M. (2004). Adaptive landscapes and dynamic equilibrium: the Spencerian contribution to twentieth-century American evolutionary biology. In A. Lustig, R. J. Richards, & M. Ruse (eds.), *Darwinian Heresies,* 131–50. Cambridge: Cambridge University Press.

Ruse, M. (2006). *Darwinism and its Discontents.* Cambridge: Cambridge University Press.

Ruse, M. (2013a). *The Cambridge Encyclopedia of Darwin and Evolution.* Cambridge: Cambridge University Press.

Ruse, M. (2013b). *Gaia in Context: Plato to Pagans.* Chicago: University of Chicago Press.

Spencer, H. (1852). A theory of population, deduced from the general law of animal fertility. *Westminster Review*, **1**, 468–501.

Spencer, H. (1857). Progress: its law and cause. *Westminster Review*, **LXVII**, 244–267.

Spencer, H. (1862). *First Principles*. London: Williams and Norgate.

Spencer, H. 1864. *Principles of Biology*. London: Williams and Norgate.

Thompson, D. W. (1917). *On Growth and Form*. Cambridge: Cambridge University Press.

Whewell, W. (1840). *The Philosophy of the Inductive Sciences*. London: Parker.

Wright, S. (1931). Evolution in Mendelian populations. *Genetics*, **16**, 97–159.

Wright, S. (1932). The roles of mutation, inbreeding, crossbreeding and selection in evolution. *Proceedings of the Sixth International Congress of Genetics*, **1**, 356–66.

13 The role of generative entrenchment and robustness in the evolution of complexity

W. C. Wimsatt

Evolutionary developmental biology is one of the most complex and articulated of the new sciences emerging out of the intersection of multiple recently separate domains. (Other new hybridizing complexes here include the new embrasure of the very large and the very small – particle physics and the big bang, characteristic of the new cosmology, and the emerging human sciences spanning from cultural anthropology and history through physical anthropology and genetics to economics and sociology.) Evolutionary developmental biology has emerged as one of the most active fields in the Darwinian sciences but the success of evolutionary developmental biology makes central the question of how to integrate an account of development into models of evolution in a general way.[1] Evolution is, after all, the evolution of developmental life-cycles, and this must be central to any account of the evolution of complexity in the biological realm. It is also important to consider the role of developmental perspectives in the intersection of cultural, scientific, and technological evolution, but this will be done below and separately.

---

[1] The distance in assumptions and the data of focus between developmental biology (or cross-phylogenetic comparative developmental genetics) and population genetics – both central to "evo-devo" as it is called, have made the two difficult to bridge. This provided the motivation for a series of modeling papers Jeffrey Schank and I have written in the last 25 years dealing with the growth and maintenance of complex systems. I won't focus on those here, except to use their results peripherally in the second part of the chapter.

Complexity and the Arrow of Time, ed. Charles H. Lineweaver, Paul C. W. Davies and Michael Ruse. Published by Cambridge University Press.
© Cambridge University Press 2013.

In this chapter, I will make some general remarks on the evolution of complexity, and then focus on one of several relevant factors or mechanisms, which I have called "generative entrenchment", and its interactions with another, robustness. Generative entrenchment has the virtue that it should apply to any evolutionary process in which there are recurrent iterative cycles of processes, and most obviously in development and life cycles. Thus it should apply to cultural, ideational, and technological evolution as well as biological evolution.[2] And it naturally tends to accumulate complexity. Robustness can act as an amplifier of this by reducing genetic load (or processes leading to degradational breakdowns, Kauffman, 1985, 1993; Lynch *et al.*, 1993). It thereby interacts with generative entrenchment in allowing the generation of much larger, more reliable, and more evolvable structures than would otherwise be possible.

I do not personally favor a search for "laws of complexity", or rather more generally for "laws" in evolution, or in the compositional sciences generally, save for special local cases. Instead, compositional systems are characterized by "sloppy, gappy regularities", which may nonetheless be quite robust and causal (Wimsatt, 1991, 2007a). This is for three reasons. Firstly, when we go up a level of organization from parts to a larger whole in the compositional sciences, even if we had an exact theory at the lower level (which we rarely do), in the move from there to the higher level, any "derivations" of regularities characteristically require approximations, exceptions, and various kinds of degeneracies (Wimsatt, 2007a, chapter 10). This rules out exact laws at that higher level. Secondly, the interesting kinds of complexity and "emergent" behavior[3] in biological and social or

---

[2] Generative entrenchment is similar to what Brian Arthur (1994) has elaborated in the technological realm as "lock in", what Rupert Riedl (1978) has baptized "burden" or what Wallace Arthur has characterized in terms of "morphogenetic trees" (Arthur, 1997). In developmental genetics, "pleiotropy" (a single gene having multiple effects) plays a similar role, although all of these concepts are characterized in slightly different ways. I summarize these approaches in Wimsatt (2007c, 2013a).

[3] Talk of emergence has figured frequently in discussions of complexity. Expanding on and analyzing the intuition "the whole is more than the sum of its parts", I have

cultural systems stem importantly from diverse interactions among differentiated parts. Regularities of behavior here would look more like engineering performance curves or digital programs – not laws. Third, the complexity increases I consider would be products, at least partially, of selection processes – often several operating simultaneously – in different modes on different units at different levels. But even selection at a single level will introduce tradeoffs (in suitably lower-dimensional cases), or more often movement in emergent directions in a higher-dimensional space.[4] Selection is characteristically a stochastic process in a high-dimensional space where the first favorable fix or "kluge" that comes along that produces a high enough fitness yield often enough is adopted. (Since it takes place in a high-dimensional space, and often with slow relaxation times, it will also likely be interrupted multiple times in midstream by other occurrent "kluges", environmental changes, or new emergent context-sensitivities, so that the system is never effectively in equilibrium.) This situation favors biasing stochastic regularities rather than law-like relations for the "satisficing"[5] performance of adaptations.[6]

tried to characterize the kind of emergence arising from features of the interaction of differentiated parts in a system in terms of kinds of failure of "aggregativity" for such systems. There are four distinct conditions that must be met for a system property to be just an "aggregate" of parts properties, and these can vary independently, producing 15 distinct ways in which a system can act non-aggregatively in exhibiting an "emergent" property. This kind of classification of modes of organization is a useful tool in analyzing the complexity of a system, and biases in simplistic theories that do not capture these relationships. See Wimsatt (2007a, chapter 12).

[4] Failure to appreciate this led to discussions (with two allele at one locus population genetic models) of individual vs group selection as if it were a "tug of war" in one dimension rather than a resultant selection vector in a higher-dimensional space. See my argument (Wimsatt, 1981b) with Williams' (1966) formulation of the problem.

[5] "Satisficing" is Herbert Simon's term for a class of heuristic and local methods for decision making in complex situations which he advanced in critique of existing unrealistic "Laplacean" models of rational decision making (see Simon, 1955, 1962 & Wimsatt, 2006a, 2000b).

[6] Thus Morange (2000) observes that the extremely high stability of the gene that seemed paradoxical and led Delbruck to phage genetics in hope of a handle to question the fundamental laws of quantum mechanics was a property of the genetic

So we should be looking for *factors* that may lead to tendencies for increasing complexity, rather than (exceptionless) law-like regularities. These complexities roughly break down into (1) trends affecting individual units, (2) trends generating complexities in or arising out of interactions between units, and (3) trends leading to or allowing articulation of smaller units into a larger unit in one, several, or many respects. In the last case, we may have conditions justifying treating the articulated units as a new higher level unit, thus moving the process up one level in that lineage.[7]

Particularly interesting are factors that may apply independently of level, so they may apply again (in a new way) with the creation of a new level, and may also (if they continue to apply at the lower level) lead to creation of entities that require multilevel or multiperspective characterization in analysis of their dynamics. One particularly important and widespread feature of multilevel systems is the emergence of robust or generic properties at the higher levels.

13.1 A SELECTIVE REVIEW OF SOME FACTORS LEADING TO COMPLEXITY INCREASE

There are several different factors that act to create trends adding to complexity, or making increased complexity possible.

(1) I believe that there is, over phylogenetic time, a robust causal tendency towards increasing complexity at any level of organization as

---

system – not of a paradoxically stable molecule. He mentions DNA repair enzymes, in addition to the redundancy implicit in the semi-conservative mode of replication, but he could have added the redundant degeneracy of the code, its relative stability to substitute amino acids with like effects on tertiary structure of proteins, and, when looking at phenotypic expression in more complex organisms, a host of features that add robustness to gene expression (Wagner, 2005.)

[7] And of course, we are generally interested in *adaptive* complexity. Richard Levins used to make this point as follows (Levins, in conversation, early 1970s): an organism is far simpler than it could be. Several hundred enzymes speed up reactions and entrain others, generating a simpler dynamics. A dung heap or a dead organism is far more complex: the enzymes are inactivated, and suddenly many thousands of interactions become dynamically important.

a result of what I have called *generative entrenchment*. This will be the focus of the second half of the chapter. But complexity cannot increase beyond limit, at any level, since there are also degenerative processes operating. If we focus on single units, it is arguable that there is a maximum of size and complexity attainable by organized systems at any given level of organization determined by the interaction of physical materials and principles and their mode or modes of organization. There are reasons why there are no people-sized single celled organisms. As evolution approaches that limit of size or performance level, the differentiation of the system into functioning subcomponents must increase at a faster rate to support marginal increases in performance (Bonner, 1965)[8], and the failure rate of the composite system, even with concurrent increases in component reliability, may increase until the size is limited by complexity catastrophe.[9] On this trajectory, escape is possible, but only by going up another level of organization, most visibly by aggregating and re-differentiating to rearticulate in such a way that the prior largest units can coordinate for redundancy and production,

---

[8] Brian Arthur (2009) calls this process "technology deepening", and tracks the number of parts in an aircraft turbojet engine through a 20-fold increase as it grows in size, efficiency, and power.

[9] Kauffman's argument in 1985 is a different kind of response to this "complexity crisis" as is Lynch *et al.* (1993) discussion of "mutational meltdown". I also have an (unpublished) rather unrealistic model dating from the 1980s of the organism as a multilevel series–parallel network of components that suggests that size can be increased subject to the constraint that reliability remains constant only if the "reserve capacity" of organs grows without bound. It is a common strategy to increase the reliability of components in larger more complex systems, and Britten (1986) documents the decrease in mutation rate (through repair enzymes and other adaptations) in the move from bacteria to metazoans. But unbounded increase in reliability for any processes is obviously impossible, so there must be a size limit for individual organisms, which can be escaped by going up a level to social organization. This line has since been exploited by Maynard Smith & Szathmary (1995). In their formulation of this transition, they view the problem as one of how to transcend competition between units, whereas I would argue that cooperative coordination is likely to play a larger role than they suppose – at least in human evolution, and likely for other social species as well. The architecture supposed for an organism in this simple model is more general than might appear, so it might be worth considering whether it could apply at higher levels, such as the ecosystem, or for human sociality (where we have multiple levels of social organization), up to the nation state and the (only too fragile) United Nations.

or if we are talking about biological organisms, production or reproduction.[10]

(2a) But how do we make that transition? There is another important source of increases in complexity, through interaction with other units, sometimes similar units (as with the emergence of human social organization), and sometimes different kinds of units closely dynamically linked in contexts with *competition or predation or parasitism* (leading to "arms races" of capacities between competitors, predator–prey, or parasite–host pairs). Thus it is plausible that the evolution of sensory and motor systems to points near the optimal limits of physical sensitivity and control, and the conjoint use of multiple complementary modalities or detection mechanisms (as with the multiple methods for distance perception available for binocular species with overlapping visual fields to increase robustness and reliability) are driven by demands of locomotion in a complex environment and the detection and capture of prey or avoidance of detection or capture by predators (Wimsatt, 1980, 2007a, chapter 10). I gather that a similar story can be told for the co-evolution of immunological capabilities and parasitic countermeasures.

(2b) Then there are other interacting systems where what appears to be happening is the *emergence of coordination*. Caporael (1995, 1997), Sterelny (2012), and increasing numbers of others argue that the main driver in human social evolution is cooperation driven through coordination rather than competition. Mackie (1996) argues that the elimination of footbinding in late nineteenth century China was accomplished through a complex coordination game affecting mate choice among Christian converts that, once a demonstrated success, spread rapidly more widely among non-Christians. Such processes are more common in culture than widely supposed in our competition-dominated society. I think Caporael, Mackie, and Sterelny are right, and that

---

[10] This schema does not cover all cases. More challenging are cases with further elaborations of complexity where we have the development of multispecies ecosystems or, even worse, ecosystem-like complexes where there are no clearly separable species among our civilized selves, organizations, and products. I believe that this situation characterizes processes of cultural evolution that have proceeded far enough to generate multiple hereditary channels with different propagation rates and bandwidths for the transmission of information horizontally. This will be discussed further below.

much of the discussion on the evolution of sociality based on competition is fundamentally misdirected. But, in either case, this has to affect the evolution of the units themselves, as well as their containing systems.

An instance of this is found in one of the most intriguing and important of biological systems, sexually reproducing species. These interactions force not only the rich interactions documented in discussions of sexual selection, but a more basic kind of selection involving genetic compatibility in breeding populations. I have argued (Wimsatt, 2006a, 2007a) that the heritability of traits and fitness in sexual species in the face of substantial genetic variability and widespread epistatic interactions have forced (selected for) a developmental architecture where most genetic substitutions don't make a difference, while many specific substitutions can nonetheless have specifiable and heritable effects generating characters for differential selection to act on. Exactly how remains a mystery, but suggestions are beginning to emerge. Azevedo *et al.* (2006) and Livnat *et al.* (2008) have elaborated this kind of insight into a "mixability" theory of the importance of sex in evolution. I believe that this sort of "cross-genomic" compatibility has been crucial in forcing the evolution of phenotypes that show significant robustness at all levels, like those pointed to by Wagner (2005). To be sure, one would have needed significant robustness at lower (and earlier) levels for sex even to be possible (one probably driven by the need to tolerate higher mutation rates among our single-cell bacterial and eukaryotic ancestors), but once sex emerged,[11] it would be a strong driver of increased "cross-platform compatibility" across other possible genomic combinations in the population.

---

[11] If "sex" was not present at the beginning. Woese (2002) argues that early evolution was dominated by substantial horizontal gene transfer for "sloppy proteins" among different quasi-proto-lineages before the emergence of mitosis and other freezing accidents that made transfers more difficult between the three emerging kingdoms of bacteria, archaea, and eukarya at the same time that the greater coding precision allowed more precise amino-acid specification, more reliable coding, and consequently large and more complex genotypes. See also Morowitz (1992) on the early fixation of different subsystems of primary metabolism.

And, indeed, forces driving standardization in hardware and software components (tracing back at least to the industrial revolution) have similar causes and similar effects. The advantages of interchangeable (replacement) parts in military hardware were an important driver to standardization and mass production in the Springfield Armory in 1828 (Hounshell, 1984; Wimsatt, 2013a), and continued demands for "platform independence" led to the Java programming language. (Here also there is seldom complete cross-compatibility, reflecting the "sloppy-gappy" generalizations characteristic for patterns of behavior in complex compositional systems.)

And what is robustness good for? It increases the reliability of a system by broadening the conditions under which it functions normally. But, in consequence, it generates neutrality or near neutrality for a wider range of genetic mutations or recombinational intersubstitutions. It thereby lowers genetic load and thus increases the complexity of systems that can be synthesized before complexity catastrophe threatens. But what generates complexity in the first place? I doubt that there is a single or a simple answer to this, but whatever congeries of factors produce locally advantageous kluges (that constitute adaptations) initially, one factor my collaborators and I have studied for many years – generative entrenchment, discussed in the next section – makes its accumulation both possible and inevitable.

The evolution of culture introduces all sorts of new kinds of complexities both in its analysis and in its characteristics (Campbell, 1974; Boyd & Richerson, 1985; Wimsatt & Griesemer, 2007). Our culture, with the various lineages of individuals, intellectual products, material artifacts, organizations (of individuals) and institutions co-developing with differentiating communication modes and channels in articulated manners seems to suggest lineages of different kinds of species. But if so, they are strange species, for none of them, even we ourselves, can any longer reproduce without the articulated resources of the others. Communication modes with more rapid transmission are co-opted to act as control loops for other communication channels, and material and informational resources criss-cross in generative and maintenance roles to aid in the production of multiple interdependent products. The convoluted complexity recalls the quip of visionary economist Kenneth Boulding that "a car is merely an organism with an exceedingly complicated sex life". (Lecture at Cornell University, fall 1974.)

Our biological paradigms provide misleadingly clear but incorrect guidance here if we look for species or units of selection at ever higher levels. We can trace multiple lineages, but most (like Boulding's car) are not self-reproducing lineages, but a complex embedded thing that reproduces with a host of other things, some of which (like human individuals) have more claim, if any do, to be the primary reproducers (Griesemer, 2000; Szathmary & Maynard Smith, 1997; Wimsatt & Griesemer, 2007). Nor are their "life-cycles" synchronous. Cars have a different life-cycle from human beings, and design practices (like standard specifications for threaded fasteners) or language dialects may outlast either. This seems an organizational mode without precedent, unless perhaps we are to take more seriously the fundamentally hybrid natures of eukaryotes and our own selves as walking ecosystems of several thousand species of our endosymbionts and parasites. But in each of these cases, we have something like an ecosystem, while in culture the dimensions and modes of interaction and co-production are such that claims to "independence" of the species types is much less than for the biological species of an ecosystem. I have tried to suggest some of the dimensions of the complexity of cultural evolution (and how it differs from biological evolution) in several places, in Wimsatt (1999), most fully (at length) in Wimsatt & Griesemer (2007), and more compactly in my critique of "memetics" as an adequate approach to cultural evolution (Wimsatt, 2010).

If we look at cultural heredity, we see other differences arising from the fact that, unlike biological organisms, which get all of their genes in a bolus at the beginning (and thus have only a single breeding population), culture is acquired by individuals sequentially as they grow up, and the reception of some traits affects both what other kinds of traits are accessible and attractive to us, and what we make of them. Moreover, as adults, we live in multiple reference groups, where we live, work, and play, and these groups are sources of information exchange – they are culturally defined breeding populations, and our inheritance draws upon and potentially affects all of them. By these standards, cultural sex is nothing if not kinky (Wimsatt, 1999, 2010). And the demographic distribution of individuals, by age, and continuing recruitment policies – the ontogeny of an organization – can have a strong effect on company culture, as Silicon Valley firms like Apple amply illustrate. Development is a crucial mediator of complexity in both biological and cultural organisms.

13.2 ADDING DEVELOPMENT TO DARWIN'S PRINCIPLES

The strategy Jeffrey Schank and I have followed (Wimsatt, 1986; Schank & Wimsatt, 1988, 2000; Wimsatt & Schank, 1988, 2004) in thinking about the evolution of complex organization is to model a very general property of developmental systems: the structure of causal dependencies relating elements in development throughout the life-cycle – in biology, from zygote to zygote. By finding a property that is general, one can construct a class of models that is widely applicable to developmental systems. These systems all have different causal dependencies of elements downstream of their various parts and processes in their development, and different magnitudes of these dependencies. One gets an evolutionary model by assessing the differential impact of these dependencies on the probable rates of evolutionary change of different elements. More downstream causal dependencies should lead to increased probability of failure with changes, and more severe failures, and should thus be more conservative in evolutionary processes. Population genetic models can reflect magnitude of dependencies (or generative entrenchment) via fitness losses of mutants that disable the traits in question. This connects these properties directly with micro-evolutionary models, and thus has advantages over models for phenotypic properties that do not have intrinsic fitness measures, which then must be posited, and seem arbitrary or special.

Any evolving system must meet what Lewontin (1970) has called "Darwin's principles":

(1) it must have descendants which differ from one another in their properties (*variation*),
(2) some of these properties must be heritable (*heritable* variation), and
(3) it must have varying causal tendencies to have descendants (heritable variation *in fitness*).

These three principles are widely advertised as core requirements for evolution. Any population of entities meeting all three conditions will undergo an evolutionary process, and nothing that fails to meet all of them can do so. Think of them as *logical or conceptual* conditions for an evolutionary process: they key into the fundamental requirements of evolution by natural selection. But these aren't the only general things one can say about evolutionary processes, or even about evolutionary processes through natural selection. Two other very general conditions reflect development's central role in the evolutionary process. No interesting evolutionary process (physical or conceptual) fails to meet them. Entities meeting the first three conditions must also:

(4) be structures generated over time: they have a developmental history (*generativity*),
(5) have parts with larger or more pervasive effects than others in that production (*differential entrenchment*).

Then different elements in the structures characteristically have downstream effects of different magnitudes. The *generative entrenchment* (GE) of an element is the magnitude of those effects in that generation or life cycle. Elements with larger degrees of GE are *generators*. This is a degree property. The generative entrenchment of an element in an evolutionary unit has deep consequences for its evolutionary fate, character, and rate of change, and that of systems impinging on it. Indeed, if earlier events in developmental trajectories show a tendency to strong conservation, the starting points will look very similar, and this phenomenon generates a trajectory that would naturally be conceived as a life cycle, so we get this important property for free. (The ways in which we are conceived and born are much more similar than the ways in which we mature and die.)[12]

So now we have five central principles of an evolutionary process; three connected with differential entrenchment, and two with development. Differential entrenchment is a high entropy (generic) property of elements of interacting structures in ensembles of such structures (Wimsatt & Schank, 1988, Wimsatt, 2001). It is possible to have a structure that depends equally (and thus symmetrically) on its different parts, but such a structure would be both very unusual and not very interesting.[13] It is a natural consequence of having differentiated parts that operation, survival, and in biology, reproduction should depend differentially upon them, and that failures of some should have more serious consequences than failures of others. So entrenchment and differential entrenchment are effectively universal in living systems (or in machines, technologies, cognitive structures, programs, or social systems of any kind).

---

[12] This requires qualification: conservation is far greater in some respects than in others. Conservation is greater right at the beginning, and remains so for crucial "housekeeping functions" like the processes that implement cell-division, and with the formation of a single-cell zygote (though the fertilized egg may differ massively in size), and again at the "phylotypic stage", but there is in some phyla substantial divergence between these (Sander, 1983; Raff, 1996). The developmental architecture must be such as to modulate entrenchment of prior states (presumably through robustness of later stages to those variations) to make this earlier divergence possible.

[13] Its parts would have to show a strong degree of causal symmetry in the consequences of their failure. The only real kind of case we are aware of like this in our experience would be strict serial circuits, like the old kind of Christmas lights that could be turned off by unscrewing any of the bulbs. Thus each bulb is necessary for a complete circuit. But this necessary symmetry also makes their behavior uninteresting, and strict serial organization for a whole organism is biologically implausible. The same could be said for strictly parallel organization – indeed it might be hard to claim that such an aggregate constituted a system.

Why should we care? Because entrenchment and differential entrenchment of different elements in the dynamical phenotype are properties that can be used at any level to predict relative evolutionary rates without knowing or explicitly using any genetics. This requires qualifications (other organizational factors such as modularity and redundancy change expected entrenchment), but it is a robust tendency and much is to be learned from the different ways in which it can fail. In areas where there is no genetic information (as in cultural evolution) we can still have predictive and explanatory evolutionary accounts using differential entrenchment, which is often also easier to detect and analyze, but if we also have genetic data, that gives us additional information about the inheritance of the traits and grounds for prediction of changes that can be used symbiotically with information about entrenchment for further predictions and analysis. Finally, differential entrenchment gives a way of tying developmental genetic information directly to fitness consequences.

While generative entrenchment says nothing about what we know as genes directly, there are some interesting conceptual connections that emerge if we step back to the period before genes were discovered. The search then was for hereditary elements that were transmitted from parents to offspring, explaining features typical of that species, as well as the individual variations that allowed us to recognize offspring of particular parents. These hereditary elements were also held to generate the distinguishing characters: Aristotle spoke of a "generative seed". But, in fact, relatively deeply entrenched features have both of these characteristics: species-typical evolutionary conservatism, and a generative role in producing characteristics of individuals. Thus things that are deeply generatively entrenched are in some ways gene-like (Wimsatt, 1981b).

The only thing genes have that generatively entrenched elements are missing is the systematic combinatorial structure that comes from the organization of genes into a common alphabet, (usually diploid) chromosomes, gametes, and genotypes – that is they are missing a (classical) genetics (Wimsatt, 1999, 2007a, chapter 11).

But must heredity be organized in this way? Consider how our many species of gut bacteria are reliably transmitted from parents to offspring through behavioral mediation (thus offspring rats eat some of their parents' feces) and re-establish their proper proportions through multispecies ecological equilibration with each other and with adaptation to their diets (the preferences for which are also transmitted from parents to offspring rats through the mothers' milk).

To visualize generative entrenchment, treat the organism's life cycle as a porous spatio-temporal sausage with characteristic events in a developmental trajectory initiating others cycling in and out.[14] Earlier events tend to have more downstream dependencies than later ones (thus greater generative entrenchment), thereby increasing their evolutionary conservatism. (See Fig. 13.1a,b.) Note from Schank's figures that ignoring cycles through the environment underestimates generative entrenchment.

This dependency can be modeled in various ways. In my original papers (Wimsatt, 1986, 1981a) I proposed a "developmental lock" which yielded an exponential decline of adaptive mutations at successively earlier stages of development. In Glassman & Wimsatt (1984), I suggested that there might be networks of "developmental locks" to allow for quasi-independent subsystems in (real) more complex cases. Rasmussen (1987) employed a scheme derived from this of series–parallel dependency systems to model the relative dependencies of the 22 then known developmental mutants in *Drosophila* – as far as I know the first attempt to construct a model for the overall architecture of development in any organism. (In spite of what it left out, and before the architecture was worked out for the role of HOX genes, it still is mostly correct.)

---

[14] Although development is normally represented as a diverging cone, we have represented it as a cylinder in order to represent in an unbiased way cases of cultural heredity. And even in biology, if we include "maternal" and stable social effects, it is not clear whether heredity and development are more cone-like or more cylinder-like.

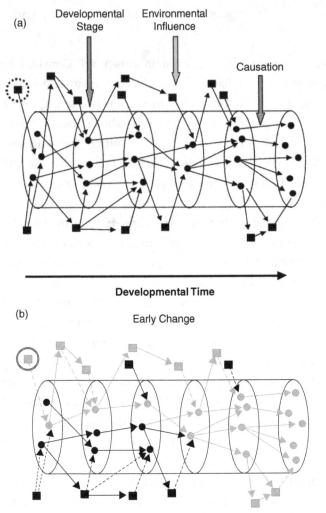

(a)

Developmental
Stage

Environmental
Influence

Causation

Developmental Time

(b)

Early Change

FIGURE 13.1 Representations of life cycle, with developmental stages and causal dependency relations (and generative entrenchment) of nodes. (a) Circular nodes are internal features of the system, square nodes are external, both characteristic features of normal development and interacting causally through the developmental stages of a phenotype or other system with a characteristic life cycle. (b) The node in the circle at upper left is changed or perturbed, with downstream consequences propagating along lighter arrows to later nodes. When they recycle out through the environment, they are dashed. (Thus, inputs from environments rapidly spread to affect all internal features, whether "innate" or otherwise (Wimsatt, 1986).) Note that ignoring outputs of internal features to the environment leads to under-estimation of both external and internal effects, since many pathways re-enter, and both can be crucial parts of a developmental program. Figure 13(a) by J. Schank and Figure 13(b) by A. Mohseni after Wimsatt and Schank (2004).

In 1985, Jeffrey Schank and I undertook a research program to model the evolution of gene control networks starting out from models introduced by Stuart Kauffman (1985) to assess how many connections representing one gene's effect on another could be preserved in large selection–mutation balance models. These network models were more general and adaptable than those I had used earlier, and thus an attractive class of objects to study.[15] Kauffman's simulations showed, and he had argued, that only relatively few genes could be maintained by selection against the degrading forces of mutation. This meant that networks of any size should be even more short-lived. Our models were similar but included the effects of generative entrenchment, which changed the outcome substantially. (Differential generative entrenchment was already present in Kauffman's networks – indeed as a generic property, but not utilized in his simulations: he was counting all losses equally.) We found strongly non-linear effects favoring preservation of the more strongly selected (more deeply entrenched) connections, confirming our expectations. So much larger networks could be maintained if they showed significant entrenchment. Surprisingly, we also found an apparent increase in the total number of connections that could be maintained relative to an otherwise equivalent model, with the same total decrements in fitness (Schank & Wimsatt, 1988; Wimsatt & Schank, 1988, 2004). This had significant effects for more molar organizational features (Schank & Wimsatt, 2001), and led us to argue (Wimsatt & Schank, 1988) that four different mechanisms and their effects could increase

---

[15] Although Kauffman's specific conclusions about the relative impotence of selection here were not robust, he has three separate times been an innovator of highly fruitful classes of modeling gene control networks: his initial move in 1969 to study randomly connected networks of Boolean automata; his later move in about 1983 to look at simpler models using directed graphs, and his third class of N-K models involving different kinds of epistatic interaction to study adaptive topographies for interactive networks have all provided productive problem spaces that have been applied far beyond his original claims, and play major roles in modern approaches to network analysis. These are reviewed in his 1993 book. Robustness and entrenchment have remained as central concerns in the analysis of gene control networks (Wagner, 2005; Wimsatt, 2007b).

the number of genes that could be maintained by selection by two to five orders of magnitude over evolutionary time.[16] This suggested that evolution should be capable of very large increases in maximum adaptive complexity. These also had other interesting and unexpected features of adaptive dynamics that we developed further in Wimsatt & Schank (2004) and Wimsatt (2013a).

Generative entrenchment can arise through simple downstream dependencies in a developing organism, but also through contributing to maintenance and stability in complex structures. Here the cyclic processes need not be life cycles but simple closed-loop feedbacks arising in the maintenance process will do, and a major role in such a feedback loop or loops can make loss very costly. This extension allows one to talk about entrenchment in developing structures that do not undergo a whole life cycle with reproduction, but which metabolize or recruit to maintain themselves. This is especially common in cultural evolution (consider firms or other organizations), but may also characterize ecological systems. The ability to apply entrenchment notions to self-maintaining organizations like business firms or professions, and cultural institutions (like languages, legal codes, or standards – SAE standards for threaded fasteners, communication protocols such as TCP/IP, or "platform independent formats" like PDF), broadens the range of application to social organizations, cultural practices, and artifacts of all sorts. More complex systems are hybrids of all three. In governments, modern manufacturing systems, and, increasingly, technological systems of all kinds, modes of dependence increase enormously and make GE a widely applicable tool in

---

[16] This was the first paper to give robust arguments for evolutionary progress at a time when that was still a very unpopular notion (which it had been at least since George Williams' book in 1966). These effects ranged from accumulation of entrenchment hierarchies in ways that converted hard selection to soft selection, to reductions in mutation rates and changes from segregation load to (much smaller) recombination load through tandem duplication of one of two heterotic alleles followed by recombination between the duplicates to make a supergene with positive epistasis.

the analysis of cultural systems. These latter things allow coordination of practices involving diverse kinds of entities and interactions and give massive increases in sustainable complexity.

A particularly important kind of entrenchment involves the emergence of "combinatorial alphabets" that can readily be used to generate a wide variety of devices that can serve further uses. These undergo explosive growth once they get to a critical mass, and generate deep entrenchment rapidly because of their immense productivity. Indeed, this was one advantage of classical genetics, one magnified with the discovery of the structure of DNA and the genetic code. Further examples would include such biological warhorses as the protein alphabet, and more recent discoveries such as the variable genes in the immune system, and the signaling molecules and cascades discussed by Kirchner & Gerhart (2005) and called "Dynamic Programming Modules" (Newman & Bhat, 2009) that affect cell assembly and differentiation in metazoans.[17] To these one must add cultural examples: the emergence of spoken words and syntax, the emergence of written icons, and then alphabetic languages, printing with movable type, and the emergence of precision adjustable machine technology capable of making a variety of standardized and interchangeable parts. These were the backbone of production in the industrial revolution, that both midwifed manufacturing of all sorts of devices and largely replaced the neighborhood blacksmith with the neighborhood hardware store.

Standardized parts become infrastructural in character (to manufacturing, replacement, and repair processes), and some infrastructural elements become the most deeply entrenched of any of our

---

[17] Kirchner & Gerhart (2005) and Newman & Bhat (2009) identify what are basically the same elements, but Newman attributes a larger role to generic physical processes in explaining how they act and should have become entrenched elements in metazoan development. Both agree, however, that they are widely found in single-celled organisms, and are co-opted for other roles as construction elements in the production of metazoan form and characteristics.

artifacts – approaching that of language itself. Water, power, gas, high-
ways, containerized shipping, telecommunications, and the Internet
all require distribution systems, earlier for material, but increasingly
for information, generating further entrenched structures and means
of communication. Here the complexity becomes almost like a physiology, and grows enormously. Were it better coordinated, it could
be for a collective intelligence, but as it is, the rich and variegated
scaffolding (Caporael *et al.*, 2013) we experience through our development and professional training (B. Wimsatt, 2013) has enormously
expanded the role for "distributed cognition" (Wilson & Clark,
2009).

Two more things deserve discussion, but cannot get space here.

(1) Both in biology and for culture, there are devices and techniques that
can allow making deeply entrenched changes easier and less risky
(though they never get either easy or risk free). These can make major
innovations possible. Not surprisingly there are more ways of doing
this for cultural elements than for biology, and this partially serves to
explain the greater rate of change and massive complexity increases in
short periods of time possible for cultural evolution. These issues are
discussed at length in the last part of Wimsatt & Griesemer (2007) and
the conditions for making deep modifications are discussed in Wimsatt
(2013a, b).

(2) I have argued (Wimsatt, 2001) that generative entrenchment naturally
produces and preserves history, and may, in biological evolution, be
the only process that systematically does so. Indeed, if this is true,
generative entrenchment is the main reason that biology is a historical science. And surely generative entrenchment is a factor in the
preservation of human culture, which means that it is also a factor
in the relevance of history to human affairs. Even when there are not
strong forces impelling the preservation of existing norms, practices,
and technologies, the simple practice of taking what is readily at hand
and modifying it induces a bias towards the past. Generative entrenchment allows progress that is at least locally cumulative, though it can
also be a source of resistance to necessary change, and of course nothing guarantees freedom from extinction. This is more important for
us today than at any time in our history. Let us hope we can make the

modifications in our collective behavior necessary to have that be a question for the longer run!

Acknowledgements

I have benefited over the years from continued interactions on this topic with my former students, Jim Griesemer, Nick Rasmussen, and especially Jeff Schank. Fig. 13.1 is redrawn from one by Schank in Wimsatt & Schank (2004), and used with his permission. I also acknowledge the textual suggestions of Barbara Wimsatt.

REFERENCES

Arthur, B. (1994). *Increasing Returns and Path-Dependence in the Economy*. Ann Arbor: Michigan University Press.

Arthur, B. (2009). *The Nature of Technology*. New York: Macmillan.

Arthur, W. (1997). *The Origin of Animal Body Plans: a Study in Evolutionary Developmental Biology*. Cambridge: Cambridge University Press.

Azevedo, R., Lohaus, R., Srinavasan, S., Dang, K., & Burch, C. (2006). Sexual reproduction selects for robustness and negative epistasis in artificial gene networks. *Nature*, **440/2**, 87–90.

Bonner, J. T. (1965). *Size and Cycle*. Princeton: Princeton University Press.

Boyd, R. & Richerson, P. (1985). *Culture and the Evolutionary Process*. Chicago: University of Chicago Press.

Britten, R. J. (1986). Rates of DNA sequence evolution differ between taxonomic groups. *Science*, **231**, 1393–1398.

Campbell, D. T. (1974). Evolutionary epistemology. In P. A. Schilpp (ed.), *The Philosophy of Karl Popper*, II. LaSalle, IL: Open Court, 413–463.

Caporael, L. (1995). Sociality: coordinating bodies, minds and groups. Psycoloquy, 6(01), Group Selection (1).www.cogsci.ecs.soton.ac.uk/cgi/psyc/newpsy?6.01.

Caporael, L. (1997). The evolution of truly social cognition: the core configuration model. *Personality and Social Psychology Review*, **1**, 276–298.

Caporael, L., Griesemer, J., & Wimsatt, W. (eds.) (2013). *Developing Scaffolding in Evolution, Cognition and Culture*. Cambridge, MA: MIT Press.

Glassman, R. B. & Wimsatt, W. C. (1984). Evolutionary advantages and limitations of early plasticity. In R. Almli and S. Finger (eds.). *Early Brain Damage*, volume I, Academic Press, 35–58.

Griesemer, J. R. (2000). Development, culture and the units of inheritance. *Philosophy of Science*, **67**, (Proceed) S348–S368.

Hounshell, D. A. (1984). *From the American System to Mass Production, 1800–1932: The Development of Manufacturing Technology in the United States.* Baltimore: The Johns Hopkins Press.

Kauffman, S. A. (1985). Self-organization, selective adaptation and its limits: a new pattern of inference in evolution and development. In D. J. Depew & Bruce H. Weber (eds.). *Evolution at a Crossroads: the New Biology and the New Philosophy of Science.* Cambridge, MA: MIT press.

Kauffman, S. (1993). *The Origins of Order.* Oxford: Oxford University Press.

Kirchner, M. & Gerhart, J. (2005). *The Plausibility of Life: Resolving Darwin's Dilemma.* New York: W. J. Norton.

Lewontin, R. C. (1970). The units of selection. *Annual Review of Ecology and Systematics,* **1**, 1–18.

Livnat, A., Papadimution, C., Dashhoft, J., & Feldman, M. V. (2008). A mixability theory for the role of sex in evolution. *PNAS,* **105**, 19803–19808.

Lynch M., Burger, R., & Gabriel, W. (1993). The mutational meltdown in asexual populations. *Journal of Heredity,* **84**: 339–344.

Mackie, G. (1996). Ending footbinding and fibulation: a convention account. *American Sociological Review,* **61**, 999–1017.

Maynard Smith, J. & Szathmary, E. (1995). *The Major Transitions in Evolution.* Oxford: Oxford University Press.

Morange, M. (2000). *Molecular Biology, a Short History.* Cambridge, MA: Harvard University Press.

Morowitz, H. (1992). *The Origins of Cellular Life.* New Haven: Yale University Press.

Newman, S. & Bhat, R. (2009). Dynamical patterning modules: a "pattern language" for development and evolution of multicellular form. *Int. J. Dev. Biol.,* **53**, 693–705.

Raff, R. (1996). *The Shape of Life: Genes, Development and the Evolution of Animal Form.* Chicago: University of Chicago Press.

Rasmussen, N. (1987). A new model of developmental constraints as applied to the *Drosophila* system. *J. Theoretical Biology,* **127**, 3, 271–301.

Riedl, R. (1978). *Order in Living Organisms: a Systems Analysis of Evolution.* New York: J. H. Wiley (translation of German original, 1975).

Sander, K. (1983). The evolution of patterning mechanisms: cleanings from insect embryogenesis and spermatogenesis. In B. C. Goodwin, N. Holder, & C. C. Wylie (eds.). *Development and Evolution,* pp. 137–159. Cambridge: Cambridge University Press.

Schank, J. C., Wimsatt, W. C. (1988). Generative entrenchment and evolution. In A. Fine and P. K. Machamer (eds.). *PSA-1986,* volume II. East Lansing: The Philosophy of Science Association, pp. 33–60.

Schank, J. C. & Wimsatt, W. C. (2001). Evolvability: adaptation, and modularity. In R. Singh, K. Krimbas, D. Paul and J. Beatty. (eds.). *Thinking about Evolution: Historical, Philosophical and Political Perspectives: Festschrift for Richard Lewontin*. Cambridge: Cambridge University Press.

Simon, H. A. (1955). A behavioral theory of rational choice. *Quarterly Journal of Economics*, **69**, 99–118.

Simon, H. A. (1962). *The Architecture of Complexity*. Philadelphia: American Philosophical Society, reprinted in *The Sciences of the Artificial*, 3rd. edn. (1996). Cambridge, MA: MIT Press.

Sterelny, K. (2012). *The Evolved Apprentice*. 2008 Jean Nicod lectures (expanded). Cambridge: MIT Press.

Szathmary, E. & Maynard Smith, J. (1997). From replicators to reproducers: the first major transitions leading to life. *Journal of Theoretical Biology*, **187**, 555–571.

Wagner, A. (2005). *Robustness and Evolvability in Living Systems*. Princeton: Princeton University Press.

Williams, G. C. (1966). *Adaptation and Natural Selection*. Princeton: Princeton University Press.

Wilson, R. & Clark, A. (2009). How to situate cognition: letting nature take its course. In M. Aydede and P. Robbins (eds.). *The Cambridge Handbook of Situated Cognition*. London: Blackwell, pp. 55–77.

Wimsatt, B. (2013). Scaffolding a career. In L. Caporael, J. Griesemer, and W. Wimsatt (eds.). *Developing Scaffolds in Evolution, Culture and Cognition*. Cambridge, MA: MIT Press.

Wimsatt, W. C. (1980). Randomness and perceived-randomness in evolutionary biology. *Synthese*, **43**, 287–329.

Wimsatt, W. C. (1981a). Robustness, reliability, and multiple-determination. In M. Brewer & B. Collins (eds.). *Scientific Inquiry and the Social Sciences*. San Francisco: Jossey-Bass, pp. 124–163. Reprinted as chapter 4 in Wimsatt (2007a).

Wimsatt, W. C. (1981b). Units of selection and the structure of the multi-level genome. In P. D. Asquith and R. N. Giere (eds.). *PSA-1980*, vol. 2. Lansing, MI: The Philosophy of Science Association, 122–183.

Wimsatt, W. C. (1986). Developmental constraints, generative entrenchment, and the innate-acquired distinction. In P. W. Bechtel (ed.) *Integrating Scientific Disciplines*. Dordrecht: Martinus-Nijhoff, pp. 185–208.

Wimsatt, W. (1991). Taming the dimensions – visualizations in science. In M. Forbes, L. Wessels, & A. Fine (eds.). *PSA-1990*, vol. 2. East Lancing, The Philosophy of Science Association, pp. 111–135.

Wimsatt, W. C. (1999). Genes, memes, and cultural inheritance. *Biology and Philosophy*, special issue on influence of R. C. Lewontin, **14**, 279–310.

Wimsatt, W. C. (2001). The generative entrenchment approach to evolving systems. In S. Oyama, R. Gray, & P. Griffiths (eds.). *Cycles of Contingency: Developmental Systems and Evolution.* Cambridge, MA: MIT Press, 219–238.

Wimsatt, W. C. (2006a). Inconsistencies, optimization and satisficing – steps towards a philosophy for limited beings: commentary on Russell Hardin. In C. Engel & L. Daston (eds.). *Is There Value in Inconsistency?* Common Goods: Law, Politics, Economics, vol. 15, Baden-Baden, Germany: Nomos Verlagsgesellschaft, pp. 201–220.

Wimsatt, W. C. (2006b). Optimization, consistency, and kluged adaptations: can maximization survive? Comment on Kachalnik *et al.,* In C. Engel & L. Daston (eds.). *Is There Value in Inconsistency?* Common Goods: Law, Politics, Economics, vol. 15, Baden-Baden, Germany: Nomos Verlagsgesellschaft, pp. 399–420.

Wimsatt, W. C. (2007a). *Re-Engineering Philosophy for Limited Beings: Piecewise Approximations to Reality.* Cambridge, MA: Harvard University Press.

Wimsatt, W. C. (2007b). On building reliable pictures with unreliable data: an evolutionary and developmental coda for the new systems biology. In F. C. Boogerd, F. J. Bruggeman, J.-H. S. Hofmeyer, & H. V. Westerhoff, (eds.). *Systems Biology: Philosophical Foundations.* Amsterdam: Reed-Elsevier, pp. 103–120.

Wimsatt, W. C. (2007c). Echoes of Haeckel? Re-entrenching development in evolution. In J. Maienschein and M. Laubichler (eds.). *Evolution and Development,* Cambridge, MA: MIT Press, pp. 309–355.

Wimsatt, W. C. (2010). Memetics does not provide a useful way of understanding cultural evolution: a developmental perspective (paired with Susan Blackmore arguing for memetics). In F. Ayala & R. Arp (eds.). *Current Controversies in Philosophy of Biology.* London: Blackwell, 273–291.

Wimsatt, W. C. (2013a). Entrenchment as a theoretical tool in evolutionary developmental biology. In A. C. Love (ed.). *Conceptual Change in Biology: Scientific and Philosophical Perspectives on Evolution and Development.* Berlin: Springer (Boston Studies in Philosophy of Science).

Wimsatt, W. C. (2013b). Evolution and the stability of functional architectures Philipe Huneman (ed.), CNRS Conference on Function and Teleology. Cambridge, MA: MIT Press.

Wimsatt, W. C. & Griesemer, J. R. (2007). Reproducing entrenchments to scaffold culture: the central role of development in cultural evolution. In R. Sansom & R. Brandon (eds.). *Integrating Evolution and Development: from Theory to Practice.* Cambridge, MA: MIT Press, pp. 228–323.

Wimsatt, W. C. & Schank, J. C. (1988). Two constraints on the evolution of complex adaptations and the means for their avoidance. In M. Nitecki. (ed.), *Evolutionary Progress*. East Lansing, MI: University of Chicago Press, 231–273.

Wimsatt, W. C. & Schank, J. C. (2004). Generative entrenchment, modularity and evolvability: when genic selection meets the whole organism. In G. Schlosser and G. Wagner (eds.). *Modularity in Evolution and Development*. Chicago: University of Chicago Press, pp. 359–394.

Woese, C. (2002). On the evolution of cells. *PNAS*, **99**(13), 8742–8747.

14 On the plurality of complexity-producing mechanisms

Philip Clayton

> When you stir your rice pudding, Septimus, the spoonful of jam
> spreads itself round making red trails like the picture of a meteor
> in my astronomical atlas. But if you stir backwards, the jam will
> not come together again. Indeed, the pudding does not notice and
> continues to turn pink just as before. Do you think this is odd?
>
> > If knowledge isn't self-knowledge it isn't doing much, mate. Is
> > the universe expanding? Is it contracting? Is it standing on one leg and
> > singing 'When Father Painted the Parlour'? Leave me out. I can expand
> > my universe without you. 'She walks into beauty, like the night of
> > cloudless climes and starry skies, and all that's best of dark and bright
> > meet in her aspect and her eyes.'
>
> > Tom Stoppard, *Arcadia*

It might seem obvious that the universe becomes more complex over
time. After all, isn't a gas cloud consisting only of hydrogen and
helium a few seconds after the big bang simpler than a cloud of hydrogen, helium, carbon, silicon, and oxygen two billion years later? Aren't
the dynamics of a system at the level of particle physics simpler than
the dynamics of a group of interacting cells?

As intuitive as these proposals are, we don't currently possess
an adequate quantitative measure of the increase or decrease in complexity either across cosmic evolution or across scientific disciplines.
In the past, many theorists presupposed that the gap between actual
entropy and maximum entropy is not permanent, that heat death
will win in the end. These theorists were eager to link complexity in
a quantitative way to this "entropy gap", so that the two would rise
and fall together. Others now argue that, in an expanding universe,
the maximum possible entropy will increase more quickly than actual

Complexity and the Arrow of Time, ed. Charles H. Lineweaver, Paul C. W. Davies
and Michael Ruse. Published by Cambridge University Press.
© Cambridge University Press 2013.

entropy, casting the "heat death" hypothesis into question. A variety of contentious topics in physics – the thermodynamics of black holes, quantum effects at macroscopic levels, and the energy density of the vacuum, among others – raise skepticism about over-quick applications of thermodynamics to cosmology. These same factors should make us cautious about construing complexity as "nothing but" the entropy gap. As a field of study, it is much richer than that.

"Complexity" is therefore not (yet?) the Grand Unified Theory of cosmic evolution, a single scientific framework adequate to describe all physical processes.[1] Complexity does not increase in a simple, straight-line fashion, any more than entropy does. Nor are all emergent properties more complex than the lower-level systems that produced them. We recognize that simpler systems combine to produce new, more complex systems. But we don't currently have sufficiently general language to describe and quantify this process. This is one of the great scientific puzzles of our day.

Despite these limitations, many theorists are tempted to take the universalizing ambitions of classical physics and to apply them to complexity theory, effectively making complexity a Theory of Everything. I share these theorists' fascination with complex systems, but for the opposite reason: the diversity rather than the unity of complex systems. Certain tempting generalizations notwithstanding, the sciences of complexity in fact study a vast variety of *distinct complexity-increasing mechanisms and systems*. On more skeptical days, I suspect that these various sciences have little more in common than the fact that they describe phenomena that we like to call "complex".

Mine is not a skepticism about the new foci of study. The topics covered in this volume, and in a host of recent conferences, are theoretically fruitful. Modeling the different types of complex systems and understanding how complex adaptive systems are able to function

---

[1] It's always perilous to name specific examples. Still, one does discern that some of the chapters in this volume tend more strongly in this direction. The chapters by Davies, Lineweaver, and Chaisson come to mind in this regard.

so successfully in different contexts is an important growth area in science. Interesting results follow from studying locally complex systems, even if they do not generalize across all of cosmic history. The detailed mathematical work is rightly creating dissatisfaction with long-accepted accounts of information and emergent properties – just to name two of the most important implications – and is giving rise to a variety of new proposals in specific sciences. Some of these are beginning to bear fruit, and others have the potential to lead to important scientific advances in the future. Perhaps some of the chapters in this volume fall into this category.

Instead, my skepticism is raised by a particular approach to complexity, which we might call the Unity Approach. The Unity Approach takes two forms. Sometimes it appears as the attempt to construct *a single science of complexity*, which is supposed to look different from any science we've ever done before, to provide a new theoretical basis to many of the specific sciences, and to unify them as fully as the "unity of science" movement in the mid-twentieth century ever dreamed. At other times this approach manifests when a scientist assumes that his particular science is adequate to explain *all* complex systems.

We should resist the Unity Approach to complexity, I suggest, not just because the evidence does not warrant its claims – though that's a good reason – *but because it obscures a deeper insight that complexity theories offer into the way that cosmic evolution works.* The goal of this chapter is to try to put words on that insight. My starting point lies in current work in complex adaptive systems theory, though I do not pause to summarize that work here. Note, however, that thinking about common features across the various complexity proposals in the literature today means working at a meta-scientific level, since no single science currently exists that encompasses these proposals. To address this particular "big question" thus requires one to draw on one's philosophical competencies – even when the goal of the exercise is, in the end, to produce better science.

In order to resist slipping into the assumption that we currently possess a science of complexity (or a unified scientific account of it), I will use the locution "a theory of complexity" throughout. When theories of complexity gesture toward common features allegedly manifested by all complex systems in the natural world, such theories are speculative, meta-scientific, and hence philosophical. This point bears repeating. In our current context, a "theory of complexity" may designate a model of complex dynamics or a theory about the nature, causes, and explanations of complexity production *in some specific empirical discipline*. If it does not do one of these two things, it is (at present) a speculative theory about analogies or similarities between theories of complexity in specific scientific domains.

The "sciences of complexity," I suggest, have at present four tasks:

- to describe and model complex dynamics;
- to produce the highest quality accounts we can currently give of complexity-producing mechanisms for some specified empirical domain and (where appropriate) to explain their adaptive function;
- to police against claims to scientific status for theories about "complexity as such"; and
- to contribute, as interest allows, to more speculative reflection on analogies between the various uses of "complexity" in the specific sciences.

It could well be that taking the focus off developing a single "science of complexity" will actually produce quicker progress on these tasks.

Several assumptions underlie this argument. Firstly, talk of complexity in evolutionary biology works only if one does not construe the increase in complexity as the primary goal of biological systems. That is, complexity-producing mechanisms (CPMs) don't *replace* differential selection as the motor of evolution. At the genetic level, effectively random variations produce phenotypical differences; those differences that increase the proportion of a given genome within a population constitute evolutionary success. Evolutionary success is, in the first place, always for the short term; it is relative to a specific fitness landscape. Because a given complex organism out-performs competitors in a given environment – say, *Tyrannosaurus* 70 MYA, or *Homo sapiens* today – does not mean that it, or complexity in general, is the telos of evolution as a whole. (We return to the question of macro-evolutionary patterns at the end.) It is not a fundamental assumption of evolutionary biology that more complex organisms are necessarily more fit for a given environment. (In fact, the more I watch Washington politics, the more ready I am to bet on the cockroaches.) And the fact that prokaryotes (the bacteria and the archaea) have continued to dominate their ecosystems for billions of years, despite the relative complexity of eukaryotic cells, shows that there is no inherent ability for the more complex cells to replace their simpler cousins. Remember that bacterial mutations are currently outperforming our best antibacterial agents in hospitals, human intelligence notwithstanding.

Secondly, I do not accept but will not here criticize four claims that one sometimes encounters: that the increase in complexity can never be quantified; that there is no net increase in complexity across evolutionary history; that we already possess a fully generalized (yet still empirical) theory of complexity; and that the dynamics of biological and cultural systems has been explained, or in principle could be explained, by a specific theory of the dynamics of physical or chemical systems.

Finally, the present chapter grows out of my recent work on complexity-producing mechanisms (CPMs) in evolutionary biology and culture, and in particular on the relationship between these two *kinds* of CPMs (Clayton, 2009). I turn to that relationship in the penultimate section but do not summarize all the arguments in the recent book. For example, for the argument to be complete, ecosystem dynamics and co-evolution in biology would have to be included, as would complex cultural and "network" dynamics such as the stock exchange and the growth of the internet. Ultimately, the complex worlds produced by the human brain – among them complexity theory itself – must surely be included among the data that a fully generalized complexity theory would have to address.

14.1 COMPARING COMPLEXITIES

I have contrasted theories of complexity in the context of specific scientific disciplines with more general debates about the nature of complexity. It doesn't take much reading in these general debates to recognize that they have a tendency to be constrained neither by data nor by well-established scientific theories. This should worry us. Hence the urgent need to develop a more disciplined approach to the current debates about complexity.

I suggest that we understand complexity, when used outside the context of a specific scientific discipline, as a theory about the relations between the more specific studies of complex systems. Theories *about theories* are called meta-level theories, or meta-theories for short. Complexity theory, understood as a meta-theory, is thus a multilevel and comparative concept.

I admit that this is a startling conclusion. After all, it's more common to talk about "the science of complexity". But we are actually only entitled to speak of the *sciences* of complexity – by which we must mean "the light shed on an (allegedly) general phenomenon by specific studies within specific disciplines". What then *is* complexity, if the analysis just given is accurate? Meta-theories are not directly scientific theories; they are not located within a specific field of science. Technically, I suppose, that makes them philosophical theories.

The very nature of many of the proposals currently being made about complexity should make clear that these proposals cannot be understood within the context of a single (existing) science. Or, put differently: if they are understood as proposals for a single new science of complexity, that science would function in a manner very similar to how Newtonian physics was supposed to function as a foundation for all science whatsoever. But a large variety of problems still have to be solved before we reach the point where such a "unified science" can be formulated.

My proposal for a meta-level theory of complexity is based on this fact. I suggest *that complexity emerges from a variety of complexity-producing mechanisms; that different mechanisms produce distinctive dynamics, different patterns of evolution over time; and that complexity may turn out, even ultimately, to be a meta-level phenomenon, that is, not a single science but a pattern, a tight or loose set of analogies, across a variety of specific empirical disciplines.*

I think there are good reasons for accepting this position. But the truth is, we don't yet know for sure. Thus the proposal is a prediction about how the study of complexity will develop over time. Since this prediction can't yet be tested, one should understand it as a sort of wager. I bet my dollar on this horse, and I think it's rational to do so. If you choose to bet on a different horse, we could have a good argument about who is more likely to be right in the long run. Still, don't forget that both of us are wagering. Until the race is run, no one knows who will be right.

Contrast complexity as a meta-level theory with, say, Newtonian physics. To have a universal Newtonian science would have meant that no meta-level scientific description of the universe is more true than the Newtonian description. In fact, every possible scientific description of the universe would have been subsumed under Newton's laws and become a special case of them. Newton's laws, together with an exhaustive description of the initial conditions (at some arbitrary point in time), would have sufficed to explain all natural phenomena past, present, and future. (Some twentieth-century variants added the requirement that one specify "bridge laws" between special sciences such as biology or psychology and the fundamental physical science. But at mid-century many physicists and philosophers took it as a matter of course that such bridge laws would soon be forthcoming (e.g., Nagel, 1961).) We now know that Newtonian physics is not a universal science in this sense; at velocities approaching c and in massive gravitational systems, relativistic effects unknown to Newton become significant. Similarly, I am betting that complexity will likewise not generalize across all systems. If this bet about the role of specific complexity-producing mechanisms is correct, it means that the dynamics of distinct systems will be irreducibly marked by the specific mechanisms or natural systems that produced them.

14.2 COMPLEXITY AND THE CULTURAL DIVIDE BETWEEN PHYSICS AND BIOLOGY

Discussions between physicists and biologists often stumble over a dichotomy that emerges at this point. The outcome is invariably a stalemate.

The one group claims that any generalizations we make – in this case, shared features that we recognize across the special sciences of complexity – will eventually evolve into mathematical models and then, ultimately, into the formulation of scientific laws. The other group (to which I belong) argues that the differences between complexity-producing mechanisms entail that no single set of laws will be adequate to the data. Consequently, we argue, generalizations about complex systems are meta-theories; the primary scientific work must lie at the level of specific complex systems – say in biology, genomics, proteomics, metabolomics; or cells, organs, organisms, ecosystems. When the level of culture is added, and even more when the meaning of specific cultural artifacts is being debated (as, for example, in the humanities), generalizations about complexity begin to include irreducibly philosophical concepts. By this point their status as philosophical theories is unmistakable.

It's not difficult to spell out the theoretical assumptions of this approach. Nature comes in levels (of organization). These levels are roughly (but only roughly) ordered by size. More complex systems are built up by aggregating stable subsystems. There are more and more "degrees of freedom".[2] Interactions across levels can cause massive changes in behavior. As a result, interactions across levels become messier and messier.

The meta-theory structure that I am proposing is meant to produce common ground so that conversation can continue even while theorists deeply disagree about fundamental features of complexity. This framework presupposes that the overarching functions of CPMs can't be fully parsed in terms of any one of the sciences of complexity in the list, nor by a single overarching science of complexity that explains all complex systems using a single mathematical model. The common features of CPMs can only be discerned through comparative studies of how they function in what are in fact highly distinct empirical fields of study.

Three features mark the resulting discussions:

- multiple specific studies of complex systems are included, and the distinct dynamics of these specific fields become part of the analysis;
- any generalizations that are drawn are adequate to the whole range of the contributing fields of study;
- the resulting theories of complexity include similarities across specific subfields, even when they are not shared by all the fields involved. There may be regional laws, but at this point there are unlikely to be laws that subsume all the fields being analyzed.

Some critics have alleged that meta-theories of this sort are inherently unstable; they must collapse into separate regional sciences, or give rise to a single overarching science of complexity, or move permanently into the sphere of philosophy. But I disagree. The creative tension produced by meta-theoretical analyses of this sort is constructive for the growth both of science and of philosophy. Think

---

[2] By "degrees of freedom" I do not mean radical or "metaphysical" freedom – the view that you can choose your next action without being strongly constrained by the sum total of causal inputs on you since birth (and before). "Degrees of freedom" refers to the possibility space that describes the sum total of possible actions for (say) an organism.

of suspended chords in music: the ear wants to quickly resolve an F-major chord where the A is replaced by a B-flat. But the suspension is musically productive; it moves us along. So likewise here. Since the Greeks, we have recognized the role of "heuristics" in empirical research.

The meta-theories framework is especially effective when one encounters "take-over bids" from one or another participant or discipline in this discussion. It helps to resist the claim that one particular area of specialization will finally subsume all the other disciplines until sufficient evidence is in hand to decide the question. Of course, focusing on such a wide range of disciplines can create tension. But it also reminds us that we are working at the frontiers of what is known (and perhaps what is knowable). Generalizations that provide a sense of "the lay of the land" – even when they move beyond what can be established scientifically – continue to be helpful when one returns to his or her home discipline.

For example, there is evidence that co-evolution can also occur between biological processes and cultural learning. In an influential book, William Durham (1991) argues that natural history reveals a dual inheritance system, involving networks of cultural as well as genetic transmission that function in continual interdependence. To take a simple example, at one time most human adults could not absorb lactose, the sugar found in milk, from non-human mammals. Today a much higher percentage of adults can absorb lactose from various milk sources, although this ability varies dramatically across populations. The evolution of the enzymes for absorbing lactose is, of course, a biological process. Yet a wide variety of cultural factors have also crucially influenced this outcome, including access to fresh milk, the practice of herding animals, the development of dairying technology, and whether drinking fresh milk is valued and encouraged within a particular culture. All these cultural practices have promoted the biological evolution of lactose tolerance in adults. Evolution of that biological capacity has in turn encouraged the further development of these particular cultural practices, in an ongoing spiral of mutual influence.

14.3 IMPLICATIONS OF THE "MULTIPLE COMPLEXITIES" APPROACH

Eight theses follow from or are suggested by the argument to this point.

(1) The definition and explanations of agents in a complex system are inseparable from the dynamics of that system. To be a unicellular organism is to be the sort of entity on which natural selection operates. Actions by biological agents of this sort only make theoretical sense when biological dynamics become an explicit part of the account.

(2) This makes organisms meta-level agents, analogous to the meta-level theory of complexity defended in this chapter. A complex organism depends on the biochemistry of its hemoglobin, the biomechanics of its physiological structures, the mechanisms that maintain homeostasis, and the neurological patterns that allow mental representations of its environment. Yet, despite the complexity of these many layers, it frequently acts as a single, unified agent in its environment. Evading a predator, it springs either left or right – and either survives or dies as a single unit.

(3) The laws of physics are necessary but not sufficient for explaining the nature and interactions of biological agents. There is an asymmetry here: Darwinian explanations must be consistent with the laws of physics, but the general laws of biology (if such exist) do not similarly constrain the motions of all physical particles. Although saber-toothed tigers, salesmen, and soccer teams consist of the same mass and energy that physicists study, their actions remain unexplained without the concepts and patterns of biology, psychology, and sociology respectively. No laws of physics are broken by these higher-order complex systems, just as no laws of physics are broken when the motion of the atoms in the rim of a wheel is explained in part by that wheel's rolling down a hill (and that motion explained, in part, by the intentions of the driver).

(4) These non-symmetrical relations between disciplines of study produce the "ladder" of the sciences. This ladder is temporally indexed and corresponds to the order of cosmic evolution.

(5) Later fields of study in the process contain all the complexities of earlier fields while adding new forms of complexity of their own. We thus encounter (what intuitively appears to be) increasing complexity as we move from physics to biology to psychology. *Pace* Descartes, mental complexity presupposes and builds from complex neural systems, even though the complexity of brains isn't enough by itself to fully explain how and why systems of mathematics or philosophy are complex in the specific ways that they are.

(6) Forms of complexity that arise later in the evolutionary process do not function independently; emergent systems rely on the complex systems available to them to further complexify a given organism or environment. The human brain, with its around 10^{11} neurons and some 10^{14} neural connections, is arguably the most complex natural system we have yet encountered in the universe. It is not surprising that it would produce mental systems complex enough not only to invent neurology but also to invent the symphony and the Italian sonnet. On the other hand, complexity does not increase in a simple, straight-line fashion. The thought of justice or world peace need not be more complex than the male's thought of the female with whom he wishes to copulate. Emergent properties are not always more complex than the systems that produced them.

(7) To pluralize complexity studies, as the CPM approach does, does not block the natural scientific study of the world but enhances it. There is more, not less, "natural piety" in emphasizing the differences among complexity-producing mechanisms, and thus the differences among the agents they produce and the actions they carry out. And there is less scientific rigor in construing all dynamical systems within the context of a single framework, à la Newton, in those cases where the differences between subsystems are essential to explaining them fully.

(8) CPMs are of course not independent. They build on one another, producing compound effects that individual CPMs could never produce. From an evolutionary perspective, there is often a selective advantage of CPMs running on CPMs. It would be unwise for a human engineer to attempt this strategy, since the outcomes would be unpredictable and he would lose control of his design. But evolution does not work by design; it works best when there is a profusion of genetic (and thus phenotypic) options that natural selection can go to work on. The method of CPMs running on CPMs has been extraordinarily effective at

developing complex organisms and ecosystems – life as we know it on this planet.

14.4 PHYSICS, BIOLOGY, CULTURE

Given the "multiple complexities" approach, one would not expect biological complexity to be identical to the complex systems that we encounter in physics. The study of evolving systems supports this conclusion. Darwinian explanations depend on selection, fitness landscapes, and the structures and functions to which they give rise. These do not appear to be dynamics that one can model in (say) thermodynamic terms alone (Kauffman & Clayton, 2006).

Systems biologists identify distinct levels in the complex systems that they study. Consider, for example, what is entailed by conceiving a cell, organ, or organism as a dynamical system. In *Closure: Emergent Organizations and their Dynamics*, Cliff Joslyn (2000, p. 71) argues "One crucial property entailed by closure is hierarchy, or the recognition of discrete levels in complex systems. Thus, the results of our discussion can be seen in the work of the hierarchy theorists... A number of systems theorists have advanced theories that recognize distinct hierarchical levels over vast ranges of physical space. Each of these levels can, in fact, be related to a level of physical closure... that is, circularly-flowing forces among a set of entities, for example among particles, cells, or galaxies... "[3] Lemke (2000, p. 100) adds "Certainly for biological systems, and probably for many others as well, the richness of their complexity derives in part from a strategy that organizes smaller units into larger ones, and these in turn into still larger units, and so on."

The way in which higher-level functions constrain phenomena at a given level (say, proteomics) represents a distinguishing feature

---

[3] The hierarchy language was rather more pronounced in the early phases of systems biology. Thus Auyang wrote in a classic text "Our sciences present the world as a hierarchical system with many branches, featuring individuals of all kinds. The individuals investigated by various sciences appear under different objective conditions... All individuals except the elementary particles are made up of smaller individuals, and most individuals are parts of larger individuals. Composition includes structures and is not merely aggregation" (Auyang, 1998, p. 40).

of biological systems. As Bernhard Palsson notes, "It is not so much the components themselves and their state that matters, contrary to the components view, but it is the state of the whole system that counts . . . We cannot construct all higher level functions from the elementary operations alone. Thus, observations and analyses of system level functions will be needed to complement the bottom-up approach. Therefore, bottom-up and top-down approaches are complementary to the analysis of the hierarchical nature of complex biological phenomena . . . There will be additional constraints and considerations that arise as we move up the hierarchy. Thus there may be measurable changes at a lower level that are inconsequential at a higher level" (Palsson, 2006, 13–14, 22–23, 284). Arguably, reconstructing how specific interactions between these different levels of analysis work lies at the very center of understanding biological systems.[4]

When one moves to the study of complexity in cultural systems, a new set of contrasts arises. Cultural complexity does not arise in the same way biological complexity does, nor can it be studied in the same way. Cultural explanations are fundamentally Lamarckian, in that acquired cultural characteristics are passed from generation to generation through social learning. They also depend (in part) on the culturally transmitted influence of new ideas and theories, the power dynamics of competing groups, the personalities of charismatic leaders, and the conscious intentions of agents. These are not dynamics that we can model in the same ways that we model the dynamics of Darwinian systems.

Of course, biology holds culture on a leash.[5] When the Shakers' beliefs cause them to stop reproducing, they won't be biologically successful. But it has turned out that the leash is much longer than the early sociobiologists thought. If a (non-reproducing) priestly

---

[4] According to the biosemiotics school (e.g., Jesper Hoffmeyer and Carl Emmeche in Copenhagen), emergent levels hierarchically interpret the levels below them and emit signs of their own, which can be interpreted across hierarchically emergent levels. See, for example, Hoffmeyer (2008).

[5] E. O. Wilson famously identified the biological influences on animal and human behavior as a "genetic leash" in *Consilience* (1998, 127–128).

class is culturally powerful enough, it may attract enough new adherents that it outperforms normally reproducing segments of society. (Some cheating is presupposed, of course.) Hence biological explanations aren't sufficient; explanations given in cultural terms do some explanatory work that can't be done without them.

The distinctive patterns of spontaneously emergent, complex order in cultural systems are studied by social psychologists, sociologists, economists, and cultural anthropologists, among others. In the study of cultural systems explanations in terms of laws play a rather smaller role. The emphasis tends to lie instead on explaining the distinctive features of individual projects and specific historical outcomes. Social scientists thus contrast "idiographic science" – the study of individual events, individuals, and epochs – with "nomothetic" or nomological (law-based) sciences (Lindlof, 2008). Succeeding in this explanatory task requires empathetic understanding (*Verstehen*), as Wilhelm Dilthey famously showed (Dilthey, 2010; Clayton, 1989). That is, one must rely on experience or "insider's knowledge" in order to formulate hypotheses and interpret data.

This is not to say that biological laws are irrelevant or that there are no laws of human behavior. But laws generally turn out to be effective for only a subset of research questions, such as predicting the behaviors of a mob or analyzing the purchasing behaviors of large groups of consumers within a given time span and culture. Only then can we make the assumption that humans will act as "ideally rational" economic agents.

It has been standard to refer to cultural patterns, artifacts, and ideas as "epigenetic". In one sense, of course, the term is unobjectionable; we are clearly dealing with phenomena above the level of genes and their direct effects. But to refer to culture as a whole as "the epigenome" is misleading in several respects. Firstly,[6] it's not clear what are the "substrates" in cultural evolution, for

---

[6] David Krakauer made this point in conversation at the complexity symposium in Phoenix, AZ in 2010.

example of writing systems (clay, papyrus, printing press, computer files, etc.). Secondly,[7] cultural creativity seems to be an exception to the absolute biological limits that elsewhere constrain species development. There are multiple limits on brain capacity, but not on mental performance (think of mathematical discovery).

More broadly, treating culture as the epigenome leads one to look for analogies with genes and their products, when in fact the cultural phenomena in question are highly *disanalogous* to the genetic transmission of information. There certainly are patterns to be discovered within cultural products such as literary styles, schools of art, the "ethos" of a culture, or different religious traditions, just as there are clearly contrasts between different mental representations. But the similarities and differences between cultural products like these require very different forms of research and testing than in physics, chemistry, or molecular biology.

14.5 COMPLEXITY AND THE "BIG QUESTIONS"

In discussions with scientists, and thus in volumes such as the present one, philosophers play two different roles: one that is oriented primarily toward the sciences, and one that turns its attention beyond them. In the first, philosophers help bring conceptual clarity to the challenges and the possible solutions within a given discipline, essentially playing a service role to the scientists who are working to move their discipline forward. The other role is to think rigorously about the questions at the borders of, and finally beyond, the existing sciences.

The meta-level theory of complexity that I have explored here includes components from both roles. In many ways it turns attention back to the study of specific complex systems, though it also encourages reflection on the more general patterns – the similarities and differences across the sciences – as a point of orientation for doing specific scientific work. But it can also turn attention in the other, more speculative direction, asking more purely philosophical

---

[7] I owe this argument to Simon Conway Morris at the same symposium.

questions about the meta-theoretical features of complexity as it is manifested in specific natural systems.

Neither the specific scientific work nor the more speculative questions are privileged, and neither should exercise hegemonic control over the other. The tragedy of the last decade (if I might editorialize for a moment) is that productive partnerships between scientists and philosophers have been rather on the decline, spotlighting instead the more domineering voices in science on the one side and the anti-scientific voices in religion on the other, each fighting to take control of the battlefield. Some now view with suspicion any collaborative efforts with philosophers, fearing a metaphysical take-over bid. The more subtle (and more productive) work – the work in which both detailed science and speculative reflection contribute as partners – is a more vulnerable form of discourse and is easily destroyed by such hostilities.

I close then by mentioning a few of the "big questions" that are raised by contemporary discussions of complexity, moving from the descriptive to the clearly metaphysical. Even if these questions can't be discussed here, it is worth noting some of the philosophical topics that the broader discussion of complexity raises when one follows its natural trajectory.

- Descriptively, it appears that this cosmos functions in such a way that complexity is increased. It is also a complexity that is non-algorithmic and open-ended. Unlike thermodynamics, we have no second law of complexity, so we do not yet know whether the increase in complexity is a necessary feature of this universe.
- The growth in complexity is of course depending on certain fundamental features of the physical universe. It is highly unlikely that complexity would increase in a universe in equilibrium, just as it is highly unlikely that complex life forms would evolve if the planet were not bathed in radiation, that is, in a far-from-thermodynamic-equilibrium state. Still, the dependence of complexity on thermodynamic conditions does not mean that thermodynamics is sufficient to explain the complexity that subsequently arises.

- The fields of study required to make sense of the full range of complexity-producing mechanisms include physical systems that one can model mathematically, chemical and biological systems, and complex neurophysiological systems such as the human brain. But they also include cultural, psychological, and intellectual systems, which are irreducible components of the complete explanation of human beings.
- Against Intelligent Design, I strongly resist using the label "science" to describe the speculations to which complexity may give rise. For example, there is no way to move up the ladder of the scientific disciplines (and the various types of complexity that they study) to produce a "scientific" proof of the existence of God as a Cosmic Designer of complexity-producing mechanisms.
- Nevertheless, one could intuitively feel that a universe of increasing complexity is the sort of universe one would expect if religious views of ultimate reality are correct. Indeed, one could correlate these religious views of reality with the universe we observe and offer metaphysical accounts of the ladder of complexity. Again, such reflection is better understood as "faith seeking understanding" rather than compelling philosophical proof – much less as a sort of metaphysical science, which is a contradiction in terms. But one can do some fairly rigorous philosophy in the attempt to turn these intuitions into philosophical arguments. Those without an ear for rigorous philosophy of this sort – and especially those who have never read it – should not be over-quick in dismissing it.
- Those with metaphysical interests begin with intuitions that are not universally shared. Intuitions should spawn deeper reflection. If one has intuitions of metaphysical purpose, one can attempt to explicate this intuition in philosophical systems (which are, I suppose, another variant of complex systems). Such systems can be either superficial or profound. Of course, one can also begin with the intuition that no such metaphysical purpose exists, and one can also develop that intuition into a broader philosophical account as well. Here also the answers given may be either superficial or profound.

The meta-level theory of complexity defended here is essentially open to discussions of this sort. Where these discussions will lead is, of course, another question.

> The ordinary-sized stuff which is our lives, the things people write poetry about – clouds – daffodils – waterfalls – what happens in

a cup of coffee when the cream goes in – these things are full of mystery, as mysterious to us as the heavens were to the Greeks.

When we have found all the meanings and lost all the mysteries, we will be alone, on an empty shore.

(Tom Stoppard, Arcadia*)*

REFERENCES

Auyang, S. (1998). Foundations of complex-system theories *Economics, Evolutionary Biology, and Statistical Physics*. Cambridge: Cambridge University Press.

Clayton, P. (1989). *Explanation from Physics to Theology: an Essay in Rationality and Religion*. New Haven, CT: Yale University Press.

Clayton, P. (2009). *In Quest of Freedom: The Emergence of Spirit in the Natural World*. Göttingen: Vandenhoeck & Ruprecht.

Dilthey, W. (2010). Understanding the Human World. In R. A. Makkreel & F. Rodi. (eds.), *Wilhelm Dilthey, selected Works*, II. Princeton, NJ: Princeton University Press.

Durham, W. H. (1991). *Coevolution: Genes, Culture, and Human Diversity*. Stanford: Stanford University Press.

Hoffmeyer, J. (2008). *Biosemiotics: an Examination into the Signs of Life and the Life of Signs*, trans. J. Hoffmeyer & D. Favareau. Scranton: University of Scranton Press.

Joslyn, C. (2000). Levels of control and closure in complex semiotic systems. In J. Chandler & G. van de Vijver (eds.), *Closure: Emergent Organizations and their Dynamics*. Annals of the New York Academy of Science Series, Volume 901. New York: New York Academy of Sciences, 67–74.

Kauffman, S. & Clayton, P. (2006). On emergence, agency, and organization. *Philosophy and Biology*, **21**, 501–21.

Lemke, J. (2000). Opening up closure: semiotics across scales. In J. Chandler & G. van de Vijver. (eds.), *Closure: Emergent Organizations and their Dynamics*, Annals of the New York Academy of Science Series, Volume **901**. New York: New York Academy of Sciences, 100–111.

Lindlof, T. R. (2008). Idiographic vs nomothetic science. In W. Donsbach (ed.), *The International Encyclopedia of Communication*, www.communicationencyclopedia.com/public/tocnode?id=g9781405131995_yr2011_chunk_g978140513199514_ss5-1.

Nagel, E. (1961). *The Structure of Science: Problems in the Logic of Scientific Explanation*. London: Routledge and Kegan Paul.

Palsson, B. (2006). *Systems Biology: Properties of Reconstructed Networks*. Cambridge: Cambridge University Press.

Wilson, E. O. (1998). *Consilience: the Unity of Knowledge*. New York: Knopf.

Index

Printed in the United States
by Baker & Taylor Publisher Services